企业级卓越人才培养（信息类专业集群）解决方案“十三五”规划教材

轻量级框架之Struts 2

天津滨海迅腾科技集团有限公司　主编

南开大学出版社
天　津

图书在版编目(CIP)数据

轻量级框架之 Struts2 / 天津滨海迅腾科技集团有限公司主编．— 天津：南开大学出版社，2017.5

ISBN 978-7-310-05328-5

Ⅰ．①轻…　Ⅱ．①天…　Ⅲ．①软件工具—程序设计　Ⅳ．①TP311.56

中国版本图书馆 CIP 数据核字（2017）第 015396 号

南开大学出版社出版发行

出版人：刘立松

地址：天津市南开区卫津路 94 号　　邮政编码：300071

营销部电话：(022)23508339　23500755

营销部传真：(022)23508542　　邮购部电话：(022)23502200

*

三河市同力彩印有限公司印刷

全国各地新华书店经销

*

2017 年 5 月第 1 版　　2017 年 5 月第 1 次印刷

260×185 毫米　16 开本　15 印张　375 千字

定价：66.00 元

如遇图书印装质量问题，请与本社营销部联系调换，电话：(022)23507125

企业级卓越人才培养（信息类专业集群）解决方案
“十三五”规划教材编写委员会

编　者：王新强　牛文峰　李树真　翟亚峰　侯梦颖

企业级卓越人才培养（信息类专业集群）解决方案简介

企业级卓越人才培养（信息类专业集群）解决方案（以下简称“解决方案”）是面向我国职业教育量身定制的应用型、技术技能型人才培养解决方案，以天津滨海迅腾科技集团技术研发为依托，联合国内职业教育领域相关行业、企业、职业院校共同研究与实践研发的科研成果。本解决方案坚持“创新产教融合协同育人，推进校企合作模式改革”的宗旨，消化吸收德国“双元制”应用型人才培养模式，深入践行“基于工作过程”的技术技能型人才培养，设立工程实践创新培养的企业化培养解决方案。在服务国家战略、京津冀教育协同发展、中国制造 2025（工业信息化）等领域培养不同层次及领域的信息化人才。为推进我国教育现代化发挥应有的作用。

该解决方案由“初、中、高级工程师”三个阶段构成，集技能型人才培养方案、专业教程、课程标准、数字资源包（标准课程包、企业项目包）、考评体系、认证体系、教学管理体系、就业管理体系等于一体。采用校企融合、产学融合、师资融合的模式在高校内共建互联网学院、软件学院、工程师培养基地的方式，开展“卓越工程师培养计划”，开设系列“卓越工程师班”，“将企业人才需求标准、企业工作流程、企业研发项目、企业考评体系、企业一线工程师、准职业人才培养体系、企业管理体系引进课堂”，充分发挥校企双方特长，推动校企、校际合作，促进区域优质资源共建共享，实现卓越人才培养目标，达到企业人才培养及招录的标准。本解决方案已在全国近二十所高校开始实施，目前已形成企业、高校、学生三方共赢格局。未来五年将努力实现在年培养能力达到万人的目标。

天津滨海迅腾科技集团是以 IT 产业为主导的高科技企业集团，总部设立在北方经济中心——天津，子公司和分支机构遍布全国近 20 个省市，集团旗下的迅腾国际、迅腾科技、迅腾网络、迅腾生物、迅腾日化分属于 IT 教育、软件研发、互联网服务、生物科技、快速消费品五大产业模块，形成了以科技为原动力的现代科技服务产业链。集团先后荣获“全国双爱双评先进单位”“天津市五一劳动奖状”“天津市政府授予 AAA 级和谐企业”“天津市文明单位”“高新技术企业”“骨干科技企业”等近百项殊荣。集团多年中自主研发天津市科技成果 2 项，具备自主知识产权的开发项目数十余项。现为国家工业和信息化部人才交流中心“全国信息化工程师”项目联合认证单位。

前　言

Struts 2 是一个基于 MVC 设计模式的 Web 应用框架，它本质上相当于一个 Servlet，在 MVC 设计模式中，Struts 2 作为控制器（Controller）来建立模型与视图的数据交互。

本书对 Struts 2 的知识进行了详细讲解，从 Struts 2 最基础的部分开始，逐步深入，可以使读者循序渐进地掌握 Struts 2 的知识。

本书共八章，分别介绍了 Struts 2 的基本概念和编写流程、Struts 2 的核心配置文件、Struts 2 转换器在项目中的作用和使用方法、如何使用 Struts 2 进行表单数据校验、Struts 2 拦截器的配置和自定义拦截器的使用、Struts 2 标签库、Struts 2 实现国际化等。

通过对本书的学习，读者可以掌握 Struts 2 核心配置文件的配置方法，Struts 2 转换器和拦截器的使用，使用 Struts 2 的标签库对代码进行简化，使用 Struts 2 实现国际化等知识，从而能够使用 Struts 2 框架进行项目开发。

本书每章均划分为学习目标、课前准备、本章简介、具体知识点讲解、小结、英语角、作业、思考题、学员回顾内容九个模块。学习目标和课前准备对本章要讲解的知识进行了简述，小结部分对本章知识进行了总结，英语角解释了本章一些术语的含义，可以使读者全面掌握本章所讲的内容。

本书由王新强、牛文峰主编，李树真、翟亚峰、侯梦颖等参与编写。王新强主编，牛文峰、李树真负责内容的规划、编排。具体分工如下：第一、二、三章由王新强、牛文峰共同编写；第四、五、六、七、八章由李树真、翟亚峰、侯梦颖共同编写。

本书示例丰富、思路清晰。通过大量的项目实践使理论知识得到了充分的应用，做到了理论与实践相结合，读者在对理论知识进行学习后可以通过配套案例来完全掌握理论部分所讲的内容。并且在每个案例之后都有相应的运行结果图，可以使读者更加直观地看到每个案例的效果，有助于对所讲知识的理解。

前　言

目　录

理论部分

上机部分

理论部分

第 1 章　Struts 2 概述

学习目标

- ✧ Struts 2 的框架架构。
- ✧ Struts 2 的控制器组件。
- ✧ Struts 1 和 Struts 2 的对比。
- ✧ 搭建 Struts 2 环境。

课前准备

- ✧ 熟悉 MVC 思想。
- ✧ 了解 Model 1 和 Model 2 。
- ✧ 熟悉 Struts 1 的基本结构。

本章简介

本章主要介绍 Struts 2 的起源以及体系结构，学习框架架构、控制器组件，了解 Struts 2 的配置文件、标签库以及 Struts 1 和 Struts 2 的不同点。

1.1　Struts 2 的起源

经过多年的发展，Struts 1 已经成为一个高度成熟的框架，不管是稳定性还是可靠性，都得到了广泛的证明。其市场占有率超过 20%，拥有丰富的开发人群，几乎已经成为事实上的工业标准。但是随着时间的流逝和技术的进步，Struts 1 的局限性也越来越多地暴露出来，并且制约了 Struts 1 的继续发展。

对于 Struts 1 框架而言，由于与 JSP/Servlet 耦合非常紧密，因而导致了一些严重的问题。首先，Struts 1 支持的表现层技术单一。其次，Struts 1 与 Servlet API 的严重耦合，使应用难于测试。最后，Struts 1 代码严重依赖于 Struts 1 API，属于侵入性框架。

从目前的技术层面上看，出现了许多与 Struts 1 竞争的视图层框架，比如 JSF、Tapestry 和 Spring MVC 等。这些框架的出现也促进了 Struts 的发展。面对大量新的 MVC 框架的兴起，Struts 1 也开始了血液的更新。目前，Struts 已经分化成了两个框架：第一个是在 Struts 1 的基础上，融合了另一个优秀的 Web 框架 WebWork 的 Struts 2。Struts 2 虽然是在 Struts 1 的基础上发展起来的，但是实质上是以 WebWork 为核心的。Struts 2 为传统的 Struts 1 注入了 WebWork 的先进的设计理念，统一了 Struts 1 和 WebWork 两个框架。Struts 1 分化出来的另外一个框架是 Shale。这个框架远远超出了 Struts 1 原有的设计思想，与原有的 Struts 1 关联很少，使用了全新的设计思想。Shale 更像一个新的框架而不是 Struts 1 的升级。本书中我们着重介绍 Struts 分化出的第一个框架 Struts 2。

当然，对于传统的 Struts 1 开发者，Struts 2 也提供了很好的兼容性，Struts 2 可与 Struts 1 有机整合，从而保证 Struts 1 开发者能平稳过渡到 Struts 2。

1.2 Struts 2 体系介绍

Struts 2 的体系与 Struts 1 体系的差别非常大，因为 Struts 2 使用了 WebWork 的设计核心。Struts 2 使用大量拦截器来处理用户请求，从而允许用户的业务逻辑控制器与 Servlet API 分离。

1.2.1 Struts 2 框架架构

从数据流程图上来看，Struts 2 与 WebWork 相差不大，Struts 2 同样使用拦截器作为处理器，以用户的业务逻辑控制器为目标，创建一个控制器代理。

控制器代理负责处理用户请求，处理用户请求时回调业务控制器的 execute() 方法，该方法的返回值将决定 Struts 2 将怎样的视图资源呈现给用户。图 1-1 显示了 Struts 2 的体系概图。

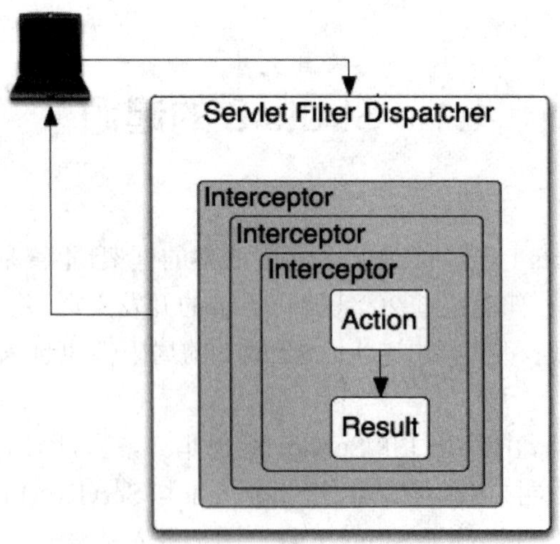

图 1-1 Struts 2 的体系概图

Struts 2 框架的大致处理流程如下：

- 浏览器发送请求，例如请求 /mypage.action 等；
- 核心控制器 FilterDispatcher（从 Struts 2.1.3 由 StrutsPrepareAndExecuteFilter 代替）根

据请求调用合适的 action；

● WebWork 的拦截器链自动对请求应用通用功能，例如 workflow、validation 或文件上传等功能；

● 回调 action 的 execute() 方法，该 execute() 方法先获取用户请求参数，然后执行某种数据库操作，既可以将数据保存到数据库，也可以从数据库中检索信息。实际上，因为 Action 只是一个控制器，它会调用业务逻辑组件来处理用户的请求；

● action 的 execute() 方法处理结果信息将被输出到浏览器中，可以是 HTML 页面、图像，也可以是 PDF 文档或者其他文档。此时支持的视图技术非常多，既支持 JSP，也支持 Velocity、FreeMarker 等模板技术。

1.2.2　Struts 2 的配置文件

当 Struts 2 创建系统的 action 代理时，需要使用 Struts 2 的配置文件。

Struts 2 的配置文件有两种方式：

● 配置 action 的 struts.xml 文件；

● 配置 Struts 2 全局属性的 struts.properties 文件。

struts.xml 文件内定义了 Struts 2 的系列 action，定义 action 时，指定该 action 的实现类，并定义该 action 处理结果与视图资源之间的映射关系。struts.xml 配置文件如示例代码 1-1 所示。

示例代码 1-1　struts.xml 配置文件的示例

```
<struts>
        <!-- Struts 2 的 Action 都必须配置在 package 里 -->
        <package name="default" extends="struts-default">
            <!-- 定义一个 logon 的 Action，实现类为 lee.Logon -->
            <action name="logon" class="lee.Logon">
                <!-- 配置 Action 返回 input 时转入 /pages/Logon.jsp 页面 -->
                <result name="input">/pages/Logon.jsp</result>
                <!-- 配置 Action 返回 cancel 时重定向到 Welcome 的 Action-->
                <result name="cancel" type="redirect-action">
                    Welcome
                </result>
                <!-- 配置 Action 返回 success 时重定向到 MainMenu 的 Action -->
                <result type="redirect-action">MainMenu</result>
                <!-- 配置 Action 返回 expired 时进入 ChangePassword 的 Action 链 -->
                <result name="expired" type="chain">
                    ChangePassword

            </result>
```

```
        </action>
        <!-- 定义 logoff 的 Action，实现类为 lee.Logoff -->
        <action name="logoff" class="lee.Logoff">
            <!-- 配置 Action 返回 success 时重定向到 MainMenu 的 Action -->
            <result type="redirect-action">MainMenu</result>
        </action>
    </package>
</struts>
```

在上面的 struts.xml 文件中，定义了两个 action。定义 action 时，不仅定义了 action 的实现类，而且在定义 action 的处理结果时，指定了多个 result，result 元素指定 execute() 方法返回值和视图资源之间的映射关系。以下配置片段：

```
<result name="cancel" type="redirect-action">Welcome</result>
```

表示当 execute() 方法返回 cancel 的字符串时，跳转到 Welcome 的 action。定义 result 元素时，可以指定两个属性：type 和 name。其中 name 指定了 execute() 方法返回的字符串，而 type 指定转向的资源类型，此处转向的资源可以是 JSP，也可以是 FreeMarker 等，甚至是另一个 action，这也是 Struts 2 可以支持多种视图技术的原因。

除此之外，Struts 2 还有一种配置 Struts 2 的方式：全局属性的 properties 文件：struts.properties。该文件的示例如下：

```
# 指定 Struts 2 处于开发状态
struts.devMode = false
# 指定当 Struts 2 配置文件改变后，Web 框架是否重新加载 Struts 2 配置文件
struts.configuration.xml.reload=true
```

如上所述，struts.properties 文件的形式是一系列的 key、value 对，它指定了 Struts 2 应用的全局属性。

1.2.3 Struts 2 的标签库

Struts 2 的标签库也是 Struts 2 的重要组成部分，Struts 2 的标签库提供了非常丰富的功能，这些标签库不仅提供了表现层数据处理，而且提供了基本的流程控制功能，还提供了国际化、Ajax 支持等功能。

通过使用 Struts 2 的标签，开发者可以最大限度地减少页面代码的书写。

JSP 页面的表单如示例代码 1-2 所示。

示例代码 1-2　传统的 HTML 标签定义表单元素示例

```
<form method="post" action="basicvalid.action">
    <!-- 下面定义三个表单域 -->
    名字:<input type="text" name="name"/><br>
    年纪:<input type="text" name="age"/><br>
    喜欢的颜色:<input type="text" name="favorite"/><br>
    <!-- 定义一个输出按钮 -->
    <input type="submit" value=" 提交 "/>
</form>
```

以上页面使用了传统 HTML 标签定义表单元素，还不具备输出校验信息的功能，Struts 2 标签的定义如示例代码 1-3 所示。

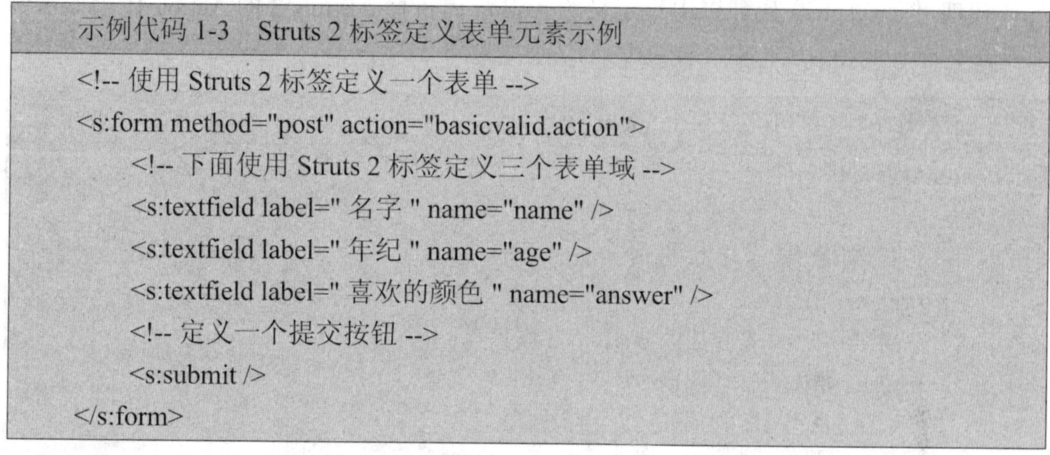

示例代码 1-3　Struts 2 标签定义表单元素示例

```
<!-- 使用 Struts 2 标签定义一个表单 -->
<s:form method="post" action="basicvalid.action">
    <!-- 下面使用 Struts 2 标签定义三个表单域 -->
    <s:textfield label=" 名字 " name="name" />
    <s:textfield label=" 年纪 " name="age" />
    <s:textfield label=" 喜欢的颜色 " name="answer" />
    <!-- 定义一个提交按钮 -->
    <s:submit />
</s:form>
```

页面代码更加简洁，而且有更简单的错误输出。图 1-2 显示的是使用 Struts 2 标签执行数据校验后的输出。

图 1-2　使用 Struts 2 标签的效果

Struts 2 的标签库的功能非常复杂，该标签库几乎可以完全替代 JSTL 的标签库。而且 Struts 2 的标签支持表达式语言，这种表达式语言支持一个强大和灵活的表达式语言：OGNL（Object Graph Navigation Language），因此功能非常强大。

1.2.4 Struts 2 的控制器组件

Struts 2 的控制器组件是 Struts 2 框架的核心，事实上，所有 MVC 框架都是以控制器组件为核心的。正如上文提到的，Struts 2 的控制器由两个部分组成：StrutsPrepareAndExecuteFilter 和业务控制器 Action。

实际上，Struts 2 应用中起作用的业务控制器不是用户定义的 Action，而是系统生成的 Action 代理，但该 Action 代理以用户定义的 Action 为目标。Struts 2 的 Action 代码如示例代码 1-4 所示。

示例代码 1-4　Struts 2 的 Action 代码示例

```
public class LoginAction
{
    // 封装用户请求参数的 username 属性
    private String username;
    // 封装用户请求参数的 password 属性
    private String password;
    // username 属性的 getter 方法
    public String getUsername()
    {
        return username;
    }
    // username 属性的 setter 方法
    public void setUsername(String username)
    {
        this.username = username;
    }
    // password 属性的 getter 方法
    public String getPassword()
    {
        return password;
    }
    // password 属性的 setter 方法
```

```
        public void setPassword(String password)
        {
            this.password = password;
        }
        // 处理用户请求的 execute() 方法
        public String execute() throws Exception
        {
            // 如果用户名为 scott，密码为 tiger，则登录成功
            if (getUsername().equals("scott") && getPassword().equals("tiger"))
            {
                return "success";
            }else{
                return "error";
            }
        }
    }
```

通过查看以上 action 代码，发现该 action 比 WebWork 中的 action 更彻底，该 action 无需实现任何父接口，无需继承任何 Struts 2 基类，该 action 类是一个完全的 POJO（普通、传统的 Java 对象），因此具有很好的复用性。

归纳起来，该 action 类有如下优势：

- action 类完全是一个 POJO，因此具有很好的代码复用性；
- action 类无需与 Servlet API 耦合，因此进行单元测试非常简单；
- action 类的 execute() 方法仅返回一个字符串作为处理结果，该处理结果可映射到任何的视图，甚至是另一个 action。

1.3　Struts 2 与 Struts 1 的对比

经过上面简要介绍，不难发现， Struts 2 确实在 Struts 1 的基础上做出了巨大的改进，是一个非常具有实用价值的 MVC 框架。下面是 Struts 1 和 Struts 2 在各方面的简要对比：

（1）action 实现类方面：Struts 1 要求 action 类继承一个抽象基类；Struts 1 的一个具体问题是使用抽象类编程而不是接口。Struts 2 的 action 类可以实现一个 action 接口，也可以实现其他接口，使可选和定制的服务成为可能。Struts 2 提供一个 ActionSupport 基类去实现常用的接口。即使 action 接口不是必须实现的，一个包含 execute() 方法的 POJO 类也可以用作 Struts 2 的 action。

（2）线程模式方面：Struts 1 的 action 是单例模式并且必须是线程安全的，因为仅有 action 的一个实例来处理所有的请求。单例策略限制了 Struts 1 的 action 能做的事，并且在开发时要

特别小心。action 资源必须是线程安全的或同步的。Struts 2 的 action 对象为每一个请求产生一个实例，因此没有线程安全问题。

（3）Servlet 依赖方面：Struts 1 的 Action 依赖于 Servlet API，因为 Struts 1 的 action 的 execute() 方法中有 HttpServletRequest 和 HttpServletResponse。Struts 2 的 action 不再依赖于 Servlet API，从而允许 action 脱离 Web 容器运行，从而降低了测试 action 的难度。当然，如果 action 需要直接访问 HttpServletRequest 和 HttpServletResponse 参数，Struts 2 的 action 仍然可以访问它们。但是，大部分时候，action 都无需直接访问 HttpServletRequest 和 HttpServletResponse，从而给开发者更多灵活的选择。

（4）可测性方面：测试 Struts 1 action 的一个主要问题是 execute() 方法依赖于 Servlet API，这使得 action 的测试要依赖于 Web 容器。为了脱离 Web 容器测试 Struts 1 的 action，必须借助于第三方扩展：Struts TestCase，该扩展下包含了系列的 Mock 对象（模拟了 HttpServletRequest 和 HttpServletResponse 对象），从而可以脱离 Web 容器测试 Struts 1 的 action 类。Struts 2 的 action 可以通过初始化、设置属性、调用方法来测试。

（5）封装请求参数：Struts 1 使用 ActionForm 对象封装用户的请求参数，所有的 ActionForm 必须继承一个基类：ActionForm。普通的 JavaBean 不能用作 ActionForm，因此，开发者必须创建大量的 ActionForm 类封装用户请求参数。虽然 Struts 1 提供了动态 ActionForm 来简化 ActionForm 的开发，但依然需要在配置文件中定义 ActionForm。Struts 2 直接使用 Action 属性来封装用户请求属性，避免了开发者需要大量开发 ActionForm 类的繁琐，实际上，这些属性还可以是包含子属性的 Rich 对象类型。如果开发者依然怀念 Struts 1 ActionForm 的模式，Struts 2 提供了 ModelDriven 模式，可以让开发者使用单独的 Model 对象来封装用户请求参数，但该 Model 对象无需继承任何 Struts 2 基类，是一个 POJO，从而降低了代码污染。

（6）表达式语言方面：Struts 1 整合了 JSTL，因此可以使用 JSTL 表达式语言。这种表达式语言有基本对象图遍历，但在对集合和索引属性的支持上则功能不强；Struts 2 可以使用 JSTL，但它整合了一种更强大和灵活的表达式语言：OGNL（Object Graph Navigation Language），因此，Struts 2 下的表达式语言功能更加强大。

（7）绑定值到视图方面：Struts 1 使用标准 JSP 机制把对象绑定到视图页面；Struts 2 使用“ValueStack”技术，使标签库能够访问值，而不需要把对象和视图页面绑定在一起。

（8）类型转换方面：Struts 1 ActionForm 属性通常都是 String 类型。Struts 1 使用 Commons-BeanUtils 进行类型转换，每个类一个转换器，转换器是不可配置的；Struts 2 使用 OGNL 进行类型转换，支持基本数据类型和常用对象之间的转换。

（9）数据校验方面：Struts 1 支持在 ActionForm 中重写 validate() 方法手动校验，或者通过整合 Commons validator 框架来完成数据校验。Struts 2 支持通过重写 validate() 方法进行校验，也支持整合 XWork 校验框架进行校验。

（10）action 执行控制方面：Struts 1 支持每一个模块对应一个请求处理（即生命周期的概念），但是模块中的所有 action 必须共享相同的生命周期。Struts 2 支持通过拦截器堆栈（Interceptor Stacks）为每一个 action 创建不同的生命周期。开发者可以根据需要创建相应堆栈，从而和不同的 action 一起使用。

1.4　WebWork 和 Struts 2 对比

从某种程度上来看，Struts 2 是 WebWork 的升级，而不是 Struts 1 的升级，甚至在 Apache 的 Struts 2 的官方文档中都提到：WebWork 到 Struts 2 是一次平滑的过渡。实际上，Struts 2 是 WebWork 2.3，从 WebWork 2.2 过渡到 Struts 2 不会比从 WebWork 2.1 到 2.2 更麻烦。

在很多方面，Struts 2 仅仅是改变了 WebWork 下的名称，因此，如果开发者具有 WebWork 的开发经验，将可以更加迅速地进入 Struts 2 的开发领域。

表 1-1 是 Struts 2 与 WebWork 在命名上存在的区别。

表 1-1　Struts 2 和 WebWork 成员名称的对应

Struts 2 成员	WebWork 成员
com.opensymphony.xwork2.*	com.opensymphony.xwork.*
org.apache.Struts2.*	com. opensymphony.webwork.*
struts.xml	xwork.xml
struts.properties	webwork.properties
Dispatcher	DispatcherUtil
org.apache.Struts2.config.Settings	com.opensymphony.webwork.config.Configuration

除此之外，Struts 2 也删除了 WebWork 中少量特性。

AroundInterceptor：Struts 2 不再支持 WebWork 中的 AroundInterceptor。如果应用程序中需要使用 AroundInterceptor，则应该自己手动导入 WebWork 中的 AroundInterceptor 类。

富文本编辑器标签：Struts 2 不再支持 WebWork 的富文本编辑器，如果应用中需要使用富文本编辑器，则应该使用 Dojo 的富文本编辑器。

IoC 容器支持：Struts 2 不再支持内建的 IoC 容器，而改为全面支持 Spring 的 IoC 容器，以 Spring 的 IoC 容器作为默认的 Object 工厂。

1.5　搭建 Struts 2 开发环境

搭建 Struts 2 开发环境时，我们一般需要做以下几个步骤的工作。

1.5.1　找到开发 Struts 2 应用需要使用到的 jar 文件

读者可以到 Apache 官方网站上下载开发 Struts 2 应用所依赖的最新版本的 jar 文件，下载地址 http://struts.apache.org/download.cgi#Struts2014，下载 struts-2.x.x-all.zip。下载完成后解压

文件，开发 Struts 2 应用需要依赖的 jar 文件在解压目录的 lib 文件夹下。不同的应用需要的 jar 包是不同的。下面给出了开发 Struts 2 程序最基本的 JAR。

（1）Struts2-core-2.x.x.jar :Struts 2 框架的核心类库。

（2）xwork-core-2.x.x.jar :XWork 类库，Struts 2 在其上构建。

（3）ognl-2.6.x.jar : 对象图导航语言（Object Graph Navigation Language），Struts 2 框架通过其读写对象的属性。

（4）freemarker-2.3.x.jar :Struts 2 的 UI 标签的模板使用 FreeMarker 编写。

（5）commons-logging-1.x.x.jar :ASF 出品的日志包，Struts 2 框架使用这个日志包来支持 Log4J 和 JDK 1.4+ 的日志记录。

（6）commons-fileupload-1.2.1.jar 文件上传组件，2.1.6 版本后必须加入此文件。

1.5.2 编写 Struts 2 的配置文件

Struts 2 默认的配置文件为 struts.xml，该文件需要存放在 WEB-INF/classes 下，该文件的配置模版如示例代码 1-5 所示。

示例代码 1-5 struts.xml 配置模板

```
<?xml version="1.0" encoding="UTF-8"?>
<!DOCTYPE struts PUBLIC
    "-//Apache Software Foundation//DTD Struts Configuration 2.3//EN"
    "http://struts.apache.org/dtds/struts-2.3.dtd">
<struts>
</struts>
```

1.5.3 在 web.xml 中加入 Struts 2 MVC 框架启动配置

在 Struts 1.x 中，Struts 框架是通过 Servlet 启动的。在 Struts 2 中，Struts 框架是通过 Filter 启动的。它在 web.xml 中的配置如示例代码 1-6 所示。

示例代码 1-6 Struts 2 在 web.xml 中的配置

```
<filter>
    <filter-name>Struts2</filter-name>
    <filter-class>org.apache.Struts2.dispatcher.ng.filter.StrutsPrepareAndExecuteFil-
ter</filter-class>
    <!-- 自从 Struts 2.1.3 以后，下面的 FilterDispatcher 已经标注为过时
    <filter-class>org.apache.Struts2.dispatcher.FilterDispatcher</filter-class>
    -->
</filter>
<filter-mapping>
    <filter-name>Struts2</filter-name>
```

```
    <url-pattern>/*</url-pattern>
</filter-mapping>
```

在 StrutsPrepareAndExecuteFilter 的 init() 方法中将会读取类路径下默认的配置文件 struts.xml 完成初始化操作。

> Struts 2 读取到 struts.xml 的内容后，以 JavaBean 形式存放在内存中，以后 Struts 2 对用户的每次请求处理将使用内存中的数据，而不是每次都读取 struts.xml 文件。

1.6 Struts 2 框架的大概处理流程

Struts 2 框架的大概处理流程如下：

（1）浏览器发送一个请求。

（2）核心控制器 FilterDispatcher 根据请求调用合适的 action。

（3）WebWork 的拦截器链自动对请求应用通用功能，如验证等。

（4）回调 action 的 execute() 方法，该 execute() 方法根据请求的参数来执行一定的操作。

（5）action 的 execute() 方法处理结果信息将被输出到浏览器中，支持多种形式的视图。

1.7 Struts 2 的优点

（1）在软件设计方面，Struts 2 不像 Struts 1 那样与 Servlet API 和 struts API 有着紧密的耦合，Struts 2 的应用可以不依赖于 Servlet API 和 struts API。Struts 2 的这种设计属于无侵入式设计，而 Struts1 则属于侵入式设计。

（2）Struts 2 提供了拦截器，利用拦截器可以进行 AOP 编程，实现如权限拦截等功能。

（3）Struts 2 提供了类型转换器，我们可以把特殊的请求参数转换成需要的类型。在 Struts1 中，如果我们要实现同样的功能，就必须向 Struts 1 的底层实现 BeanUtil 注册类型转换器才行。

（4）Struts 2 提供支持多种表现层技术，如 JSP、freeMarker、Velocity 等。

（5）Struts 2 的输入校验可以对指定方法进行校验，解决了 Struts 1 的长久之痛。

（6）提供了全局范围、包范围和 action 范围的国际化资源文件管理实现。

1.8 第一个 Struts 2 项目

在默认的配置文件 struts.xml 中加入如示例代码 1-7 所示的配置。

示例代码 1-7 struts.xml 中的配置

```
<?xml version="1.0" encoding="UTF-8"?>
<!DOCTYPE struts PUBLIC
    "-//Apache Software Foundation//DTD Struts Configuration 2.3//EN"
    "http://struts.apache.org/dtds/struts-2.3.dtd">
<struts>
    <package name="test" namespace="/test" extends="struts-default">
        <action name="helloworld"
            class="com.xtgj.struts2.action.HelloWorldAction" method="execute">
            <result name="success">/WEB-INF/page/hello.jsp</result>
        </action>
    </package>
</struts>
```

编写 HelloWorldAction，代码如示例代码 1-8 所示。

示例代码 1-8 HelloWorldAction 代码

```
package com.xtgj.struts2.action;

public class HelloWorldAction {
    private String message;

    public String getMessage() {
        return message;
    }

    public void setMessage(String message) {
        this.message = message;
    }

    public String execute() {
        this.message = " 我的第一个 struts2 应用 ";
```

```
            return "success";
        }

    }
```

在 web.xml 中添加如示例代码 1-9 所示的配置。

示例代码 1-9　web.xml 中的配置

```
<?xml version="1.0" encoding="UTF-8"?>
<web-app version="2.4" xmlns="http://java.sun.com/xml/ns/j2ee"
    xmlns:xsi="http://www.w3.org/2001/XMLSchema-instance"
    xsi:schemaLocation="http://java.sun.com/xml/ns/j2ee
    http://java.sun.com/xml/ns/j2ee/web-app_2_4.xsd">
    <filter>
        <filter-name>struts2</filter-name>
        <filter-class>
            org.apache.struts2.dispatcher.ng.filter.StrutsPrepareAndExecuteFilter
        </filter-class>
    </filter>
    <filter-mapping>
        <filter-name>struts2</filter-name>
        <url-pattern>/*</url-pattern>
    </filter-mapping>
    <welcome-file-list>
        <welcome-file>index.jsp</welcome-file>
    </welcome-file-list>
</web-app>
```

在地址栏中输入“http://localhost:8080/Struts2_Chapter01/test/helloworld.action”，将得到如图 1-3 所示的运行结果。

图 1-3　运行结果

1.9 小结

✓ Web 应用的开发历史。

✓ Model 1 和 Model 2 的简要模型和特征。

✓ MVC 模式的主要策略和主要优势。

✓ 常用的 MVC 框架，包括 JSF、Tapestry 和 Spring MVC，以及这些框架的基本知识和相关特征。

✓ Struts 2 的两个前身：Struts 1 和 WebWork，以及这两个框架的架构和主要特征。

✓ Struts 2 起源的介绍。

✓ Struts 2 框架的体系，包括 Struts 2 框架的架构、标签库、控制器组件等。

✓ 对 Struts 1 和 Struts 2 的相关方面进行比较。

✓ 搭建 Struts 2 开发环境。

1.10 英语角

redirect 重定向
model 模型
filter 过滤
execute 执行

1.11 作业

简述 Struts 2 的工作流程。

1.12 思考题

根据已经学过的 Struts 1 的有关内容，对比一下和 Struts 2 有哪些不同，Struts 2 有哪些方面较之 Struts 1 的功能要强。

1.13　学员回顾内容

1.MVC 思想。

2.Struts 2 的工作流程。

3.Struts 1 和 Struts 2 对比。

第 2 章　Struts 2 基础

学习目标

- ✧ 为 Web 应用增加 Struts 2 支持。
- ✧ Struts 2 框架的 MVC 组件。
- ✧ Struts 2 框架的流程。
- ✧ 通过 web.xml 文件加载 Struts 2 框架。

课前准备

- ✧ Struts 1 框架的基本知识。
- ✧ 使用 Struts 1 框架开发 Web 应用。
- ✧ 在 MyEclipse 中整合 Tomcat。

本章简介

本章主要介绍 Struts 2 的开发环境配置，学习基于 Struts 2 的 Web 应用开发的基本方法，了解 Struts 2 的基本特性。

2.1　在 MyEclipse 中开发 Struts 2

2.1.1　从用户请求开始

Struts 2 支持大部分视图技术，当然也支持最传统的 JSP 视图技术，本应用将使用最基本的视图技术：JSP 技术。当用户需要登录系统时，用户需要一个简单的表单提交页面，这个表单提交页面包含了两个表单域：用户名和密码。

下面是一个最简单的表单提交页面，该页面的表单内仅包含两个表单域，甚至没有任何动态内容，实际上，整个页面完全可以是一个静态 HTML 页面。但考虑到需要在该页面后增加动态内容，因此依然将该页面以“.jsp”为后缀保存。用户请求登录的 JSP 页面如示例代码 2-1 所示。

示例代码 2-1　用户请求登录的 JSP 页面代码

```
<%@ page language="java" pageEncoding="UTF-8"%>
<html>
	<head>
		<title> 登录页面 </title>
	</head>
	<body>
		<!-- 提交请求参数的表单 -->
		<form action="login.action" method="post">
			<table align="center">
				<caption>
					<h3>
						用户登录
					</h3>
				</caption>
				<tr>
					<!-- 用户名的表单域 -->
					<td>
						用户名：
						<input type="text" name="username" />
					</td>
				</tr>
				<tr>
					<!-- 密码的表单域 -->
					<td>
						密    码：
						<input type="password" name="password" />
					</td>
				</tr>
				<tr align="center">
					<td colspan="2">
						<input type="submit" value=" 登录 " />
						<input type="reset" value=" 重填 " />
					</td>
				</tr>
			</table>
		</form>
```

```
    </body>
</html>
```

如上文所述，该页面没有包含任何的动态内容，完全是一个静态的 HTML 页面。但我们注意到该表单的 action 属性：login.action，这个 action 属性比较特殊，它不是一个普通的 Servlet，也不是一个动态 JSP 页面。可能读者已经猜到了，当表单提交给 login.action 时，Struts 2 的 StrutsPrepareAndExecuteFilter 将自动起作用，将用户请求转发到对应的 action。

该页面就是一个基本的 HTML 页面，在浏览器中浏览该页面，看到如图 2-1 所示的界面。

图 2-1 用户登录界面

整个页面就是一个标准的 HTML 页面，整个单独的页面还没有任何与用户交互的能力。下面我们开始动手创建一个 Struts 2 的 Web 应用。

2.1.2 创建 Web 应用

为了让读者更加清楚 Struts 2 应用的核心，下面将“徒手”建立一个 Struts 2 应用。

建立一个 Web 应用请按如下步骤进行：

（1）在任意目录新建一个文件夹，并以该文件夹建立一个 Web 应用。

（2）在第 1 步所建的文件夹内建一个 WEB-INF 文件夹，进入 Tomcat，或任何 Web 容器内，找到一个 Web 应用，将 Web 应用的 WEB-INF 下的 web.xml 文件复制到第 2 步所建的 WEB-INF 文件夹下。

（3）修改复制的 web.xml 文件，将该文件修改成只有一个根元素的 XML 文件，修改后的 web.xml 文件代码如示例代码 2-2 所示。

示例代码 2-2 修改后的 web.xml 文件

```
<?xml version="1.0" encoding="UTF-8"?>
<web-app version="2.4" xmlns="http://java.sun.com/xml/ns/j2ee"
    xmlns:xsi="http://www.w3.org/2001/XMLSchema-instance"
    xsi:schemaLocation="http://java.sun.com/xml/ns/j2ee
    http://java.sun.com/xml/ns/j2ee/web-app_2_4.xsd">
```

```
        <filter>
            <filter-name>struts2</filter-name>
            <filter-class>
                org.apache.struts2.dispatcher.ng.filter.StrutsPrepareAndExecuteFilter
            </filter-class>
        </filter>
        <filter-mapping>
            <filter-name>struts2</filter-name>
            <url-pattern>/*</url-pattern>
        </filter-mapping>
        <welcome-file-list>
            <welcome-file>index.jsp</welcome-file>
        </welcome-file-list>
    </web-app>
```

在第 2 步所建的 WEB-INF 路径下，新建两个文件夹：classes 和 lib，它们分别用于保存单个 *.class 文件和 jar 文件。

经过上述步骤，已经建立了一个空 Web 应用。将该 Web 应用复制到 Tomcat 的 webapps 路径下，该 Web 应用将可以自动部署在 Tomcat 中。

将上一步所定义的 JSP 页面文件复制到第 1 步所建的文件夹下，该 JSP 页面将成为该 Web 应用的一个页面。该 Web 将有如图 2-2 所示的文件结构。

```
WebRoot
  META-INF
  WEB-INF
  login.jsp
```

图 2-2　web 目录结构

login.jsp 是该 Web 应用下 JSP 页面的名字，也可修改。其他文件夹、配置文件都不可修改。

启动 Tomcat，在浏览器中输入上一步定义的 JSP 页面的地址，将看到如图 2-1 所示的页面。

为了给 Web 应用增加 Struts 2 功能，只需要将 Struts 2 安装到 Web 应用中即可。具体步骤参考上一章中介绍的 Struts 2 开发环境的搭建：

（1）找到开发 Struts 2 应用需要使用到的 jar 文件。

（2）编写 Struts 2 的配置文件。

（3）在 web.xml 中加入 Struts 2 MVC 框架启动配置。

2.1.3　实现控制器类

Struts 2 下的控制器不再像 Struts 1 下的控制器，需要继承一个 Action 父类，甚至可以无需实现任何接口，Struts 2 的控制器就是一个普通的 POJO。

实际上，Struts 2 的 action 就是一个包含 execute() 方法的普通 Java 类，该类里包含的多个属性用于封装用户的请求参数。处理用户请求的 Action 类的代码如示例代码 2-3 所示。

示例代码 2-3　处理用户请求的 Action 类的代码

```
package com.xtgj.struts2.chapter02;
//Struts 2 的 Action 类就是一个普通的 Java 类
public class LoginAction {
    // 下面是 Action 内用于封装用户请求参数的两个属性
    private String username;
    private String password;
    // username 属性对应的 getter 方法
    public String getUsername() {
        return username;
    }
    // username 属性对应的 setter 方法
    public void setUsername(String username) {
        this.username = username;
    }
    // password 属性对应的 getter 方法
    public String getPassword() {
        return password;
    }
    // password 属性对应的 setter 方法
    public void setPassword(String password) {
        this.password = password;
    }
    // 处理用户请求的 execute() 方法
    public String execute() throws Exception {
        // 当用户请求参数的 username 等于 scott，密码请求参数为 tiger 时，返回
success 字符串
        // 否则返回 error 字符串
        if (getUsername().equals("scott") && getPassword().equals("tiger")) {
            return "success";
        } else {
            return "error";
        }
    }
}
```

上述的 action 类是一个再普通不过的 Java 类，该类里定义了两个属性：username 和 password，并为这两个属性提供了对应的 setter 和 getter 方法。除此之外，该 action 类里还包含了一个无参数的 execute() 方法——这大概也是 action 类与 POJO 唯一的差别。实际上，这个 execute() 方法依然是一个很普通的方法，既不与 Servlet API 耦合，也不与 Struts 2 API 耦合。

2.1.4 配置 action

前文定义了 Struts 2 的 action，但该 action 还未配置在 Web 应用中，不能处理用户请求。为了让该 action 能处理用户请求，还需要将该 action 配置在 struts.xml 文件中。

如上文所述，struts.xml 文件应该放在 classes 路径下，该文件主要放置 Struts 2 的 action 定义。定义 action 时，除了需要指定该 action 的实现类外，还需要定义 action 处理结果和资源之间的映射关系。示例代码 2-4 是该应用的 struts.xml 文件的代码。

示例代码 2-4 struts.xml 代码

```
<?xml version="1.0" encoding="UTF-8" ?>
<!-- 指定 Struts 2 配置文件的 DTD 信息 -->
<!DOCTYPE struts PUBLIC
    "-//Apache Software Foundation//DTD Struts Configuration 2.3//EN"
    "http://struts.apache.org/dtds/struts-2.3.dtd">
<!-- struts 是 Struts 2 配置文件的根元素 -->
<struts>
    <!-- Struts 2 的 Action 必须放在指定的包空间下定义 -->
    <package name="xtgj" extends="struts-default">
        <!-- 定义 login 的 Action，该 Action 的实现类为 com.xtgj.ch02.Action 类 -->
        <action name="login" class="com.xtgj.struts2.chapter02.LoginAction">
            <!-- 定义处理结果和资源之间映射关系。 -->
            <result name="error">/error.jsp</result>
            <result name="success">/welcome.jsp</result>
        </action>
    </package>
</struts>
```

上述映射文件定义了 name 为 login 的 action，即该 action 将负责处理向 login.action URL 请求的客户端请求。该 action 将调用自身的 execute() 方法处理用户请求，如果 execute() 方法返回 success 字符串，请求将被转发到 welcome.jsp 页面；如果 execute() 方法返回 error 字符串，则请求被转发到 error.jsp 页面。

2.1.5 增加视图资源完成应用

经过上述步骤，再为该 Web 应用增加两个 JSP 文件，分别是 error.jsp 页面和 welcome.jsp 页面，将这两个 JSP 页面文件放在 Web 应用的根路径下（与 WEB-INF 在同一个文件夹下），这个最简单的 Struts 2 应用马上就可以运行了。

这两个 JSP 页面文件是更简单的页面，它们只包含了简单的提示信息。其中 welcome.jsp 页面的代码如示例代码 2-5 所示。

示例代码 2-5 welcome.jsp 页面的代码

```
<%@ page language="java" pageEncoding="UTF-8"%>
<html>
    <head>
        <title> 成功页面 </title>
    </head>
    <body>
        <center> 您已经登录 !</center>
    </body>
</html>
```

上述的页面就是一个普通的 HTML 页面，登录失败后进入的 error.jsp 页面也与此类似。

在图 2-1 所示页面的"用户名"输入框中输入 scott，在"密码"输入框中输入 tiger，页面将进入 welcome.jsp 页面，将看到如图 2-3 所示的页面。

上述处理流程可以简化为如下的流程：用户输入两个参数，即 username 和 password，然后向 login.action 发送请求，该请求被 StrutsPrepareAndExecuteFilter 转发给 LoginAction 处理，如果 LoginAction 处理用户请求返回 success 字符串，则返回给用户 welcome.jsp 页面，如图 2-3 所示；如果返回 error 字符串，则返回给用户 error.jsp 页面，如图 2-4 所示。

图 2-3 登录成功页面

图 2-4　登录失败页面

error.jsp 页面的代码如示例代码 2-6 所示。

示例代码 2-6　error.jsp 页面的代码

```
<%@ page language="java" pageEncoding="UTF-8"%>
<html>
    <head>
        <title> 失败页面 </title>
    </head>
    <body>
        <center> 您登录失败 !</center>
    </body>
</html>
```

2.1.6　改进控制器实现 Action 接口

通过前文介绍，读者已经可以完成简单的 Struts 2 的基本应用了，但还可以进一步改进应用的 action 类，例如该 action 类可以通过实现 action 接口，利用该接口的优势。前文中应用的 Action 类没有与 JavaBean 交互，没有将业务逻辑操作的结果显示给客户端。

实现 Struts 2 的 Action 接口看似没有太大的好处，仅会污染该 action 的实现类。事实上，实现 action 接口可以帮助开发者更好地实现 action 类。action 接口的定义如示例代码 2-7 所示。

示例代码 2-7　Action 接口的定义

```
package com.xtgj.struts2.chapter02;

public interface Action{
    // 下面定义了 5 个字符串常量
    public static final String SUCCESS = "success";
    public static final String NONE = "none";
    public static final String ERROR = "error";
```

```
        public static final String INPUT = "input";
        public static final String LOGIN = "login";
        // 定义处理用户请求的 execute 抽象方法
        public String execute() throws Exception;
}
```

在上述的 Action 代码中，我们发现该 Action 接口里已经定义了 5 个标准字符串常量：SUCCESS、NONE、ERROR、INPUT 和 LOGIN，它们可以简化 execute() 方法的返回值，并可以使用 execute() 方法的返回值标准化。如果处理成功，则返回 SUCCESS 常量，避免直接返回一个 success 字符串（程序中应该尽量避免直接返回数字常量、字符串常量等）。

因此，借助于上面的 action 接口，可将原来的 action 类代码修改为如示例代码 2-8 所示

示例代码 2-8　实现 Action 接口的 LoginAction 类代码

```
package com.xtgj.struts2.chapter02;

// 实现 Action 接口来实现 Struts 2 的 Action 类
public class LoginAction implements Action {
        // 下面是 Action 内用于封装用户请求参数的两个属性
        private String username;
        private String password;

        // username 属性对应的 getter 方法
        public String getUsername() {
                return username;
        }

        // username 属性对应的 setter 方法
        public void setUsername(String username) {
                this.username = username;
        }

        // password 属性对应的 getter 方法
        public String getPassword() {
                return password;
        }
        // password 属性对应的 setter 方法
        public void setPassword(String password) {
                this.password = password;
```

```
        }
        // 处理用户请求的 execute() 方法
        public String execute() throws Exception {
            // 当用户请求参数的 username 等于 scott，密码请求参数为 tiger 时，返
回 success 字符串
            // 否则返回 error 的字符串
            if (getUsername().equals("scott") && getPassword().equals("tiger")) {
                return SUCCESS;
            } else {
                return ERROR;
            }
        }
    }
```

对比前文中的 action 和此处的 action 实现类，发现两个 action 类的代码基本相似，后面的 action 类实现了 Action 接口，故 action 类的 execute() 方法可以返回 Action 接口里的字符串常量。

以上是一个 Struts 2 的应用实例，从 action 类基本流程控制讲起，详细介绍了如何开发一个 Struts 2 应用。

2.2　Struts 2 的基本配置

前文大致介绍了 Struts 2 框架的基本内容，但这些基本内容都必须建立在 Struts 2 的配置文件基础之上，这些配置文件的配置信息也是 Struts 2 应用的核心部分。

2.2.1　配置 web.xml 文件

任何 MVC 框架都需要与 Web 应用整合，这就不得不借助于 web.xml 文件，只有配置在 web.xml 文件中 Servlet 才会被应用加载。

通常，所有的 MVC 框架都需要 Web 应用加载一个核心控制器，对于 Struts 2 框架而言，需要加载 StrutsPrepareAndExecuteFilter，只要 Web 应用负责加载 StrutsPrepareAndExecuteFilter，StrutsPrepareAndExecuteFilter 将会加载应用的 Struts 2 框架。

因为 Struts 2 将核心控制器设计成 Filter，而不是一个普通 Servlet。故为了让 Web 应用加载 StrutsPrepare- AndExecuteFilter，只需要在 web.xml 文件中配置 StrutsPrepareAndExecuteFilterFilter 即可。

配置 StrutsPrepareAndExecuteFilter 的代码片段如示例代码 2-9 所示。

示例代码 2-9　web.xml 文件中配置 Filter 的代码

```
<!-- 配置 Struts 2 框架的核心 Filter -->
<filter>
        <!-- 配置 Struts 2 核心 Filter 的名字 -->
        <filter-name>struts2</filter-name>
        <!-- 配置 Struts 2 核心 Filter 的实现类 -->
        <filter-class>
    org.apache.struts2.dispatcher.ng.filter.StrutsPrepareAndExecuteFilter
        </filter-class>
</filter>
```

在 web.xml 文件中配置了该 Filter，还需要配置该 Filter 拦截的 URL。通常，我们让该 Filter 拦截所有的用户请求，因此使用通配符来配置该 Filter 拦截的 URL。

配置该 Filter 拦截 URL 的配置片段如示例代码 2-10 所示。

示例代码 2-10　StrutsPrepareAndExecuteFilter 拦截 URL 的配置片段

```
<!-- 配置 Filter 拦截的 URL -->
<filter-mapping>
        <!-- 配置 Struts 2 的核心 StrutsPrepareAndExecuteFilter 拦截所有用户请求 -->
        <filter-name>struts2</filter-name>
        <url-pattern>/*</url-pattern>
</filter-mapping>
```

配置了 Struts 2 的核心 StrutsPrepareAndExecuteFilter 后，基本完成了 Struts 2 在 web.xml 文件中的配置。

如果 Web 应用使用了 Servlet 2.3 以前的规范，因为 Web 应用不会自动加载 Struts 2 框架的标签文件，因此必须在 web.xml 文件中配置加载 Struts 2 标签库。

如果 Web 应用使用 Servlet 2.4 以上的规范，则无需在 web.xml 文件中配置标签库定义，因为 Servlet 2.4 规范会自动加载标签库定义文件。因为本书采用 Servlet 2.4，所以在此不做配置。

Struts 2 的标签库定义文件包含在 struts2-core-2.3.15.3.jar 文件里，在 struts2-core-2.3.15.3.jar 文件的 META-INF 路径下，包含了一个 struts-tag.tld 文件，这个文件就是 Struts 2 的标签库定义文件，Servlet 2.4 规范会自动加载该标签库文件。

对于 Servlet 2.4 以上的规范，Web 应用自动加载该标签库定义文件。加载 struts-tag.tld 标签库定义文件时，该文件的开始部分包含如示例代码 2-11 所示的代码片段。

示例代码 2-11　struts-tag.tld 标签库定义代码片段

```
<taglib>
        <!-- 定义标签库的版本 -->
        <tlib-version>2.2.3</tlib-version>
        <!-- 定义标签库所需的 JSP 版 -->
        <jsp-version>1.2</jsp-version>
        <short-name>s</short-name>
        <!-- 定义 Struts 2 标签库的 URI -->
        <uri>/struts-tags</uri>
        ...
</taglib>
```

因为该文件中已经定义了该标签库的 URI：/struts-tags，这就避免了在 web.xml 文件中重新定义 Struts 2 标签库文件的 URI。

2.2.2　struts.xml 配置文件

Struts 2 框架的核心配置文件就是 struts.xml 配置文件，该文件主要负责管理 Struts 2 框架的业务控制器 Action。

在默认情况下，Struts 2 框架将自动加载放在 WEB-INF/classes 路径下的 struts.xml 文件。在大部分应用里，随着应用规模的增加，系统中 action 数量也大量增加，导致 struts.xml 配置文件变得非常臃肿。

为了避免 struts.xml 文件过于庞大、臃肿，提高 struts.xml 文件的可读性，我们可以将一个 struts.xml 配置文件分解成多个配置文件，然后在 struts.xml 文件中包含其他配置文件。

示例代码 2-12 所示的 struts.xml 文件就通过 include 手动导入了一个配置文件：struts-part1.xml 文件，通过这种方式，就可以将 Struts 2 的 Action 按模块配置在多个配置文件中。

示例代码 2-12　Struts 2 的 Action 按模块配置在多个配置文件示例

```
<?xml version="1.0" encoding="UTF-8" ?>
<!-- 指定 Struts 2 配置文件的 DTD 信息 -->
<!DOCTYPE struts PUBLIC
        "-//Apache Software Foundation//DTD Struts Configuration 2.3//EN"
        "http://struts.apache.org/dtds/struts-2.3.dtd">
<!-- 下面是 Struts 2 配置文件的根元素 -->
<struts>
        <!-- 通过 include 元素导入其他配置文件 -->
        <include file="struts-part1.xml" />
        ...
</struts>
```

通过这种方式，Struts 2 提供了一种模块化的方式来管理 struts.xml 配置文件。

除此之外，Struts 2 还提供了一种插件式的方式来管理配置文件。用 WinRAR 等解压缩软件打开 struts2-core-2.3.15.3.jar 文件，看到如图 2-5 所示的文件结构，其中有一个 struts-default.xml 文件。

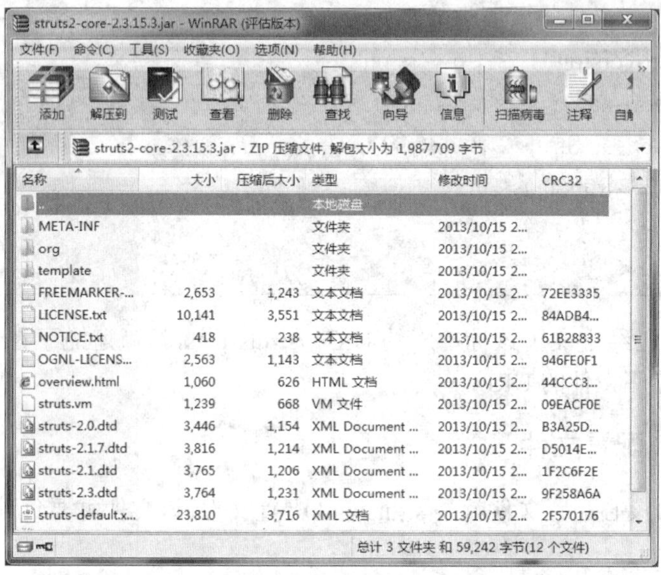

图 2-5 struts2-core-2.3.15.3.jar 的文件结构

查看 struts-default.xml 文件，该文件如示例代码 2-13 所示。

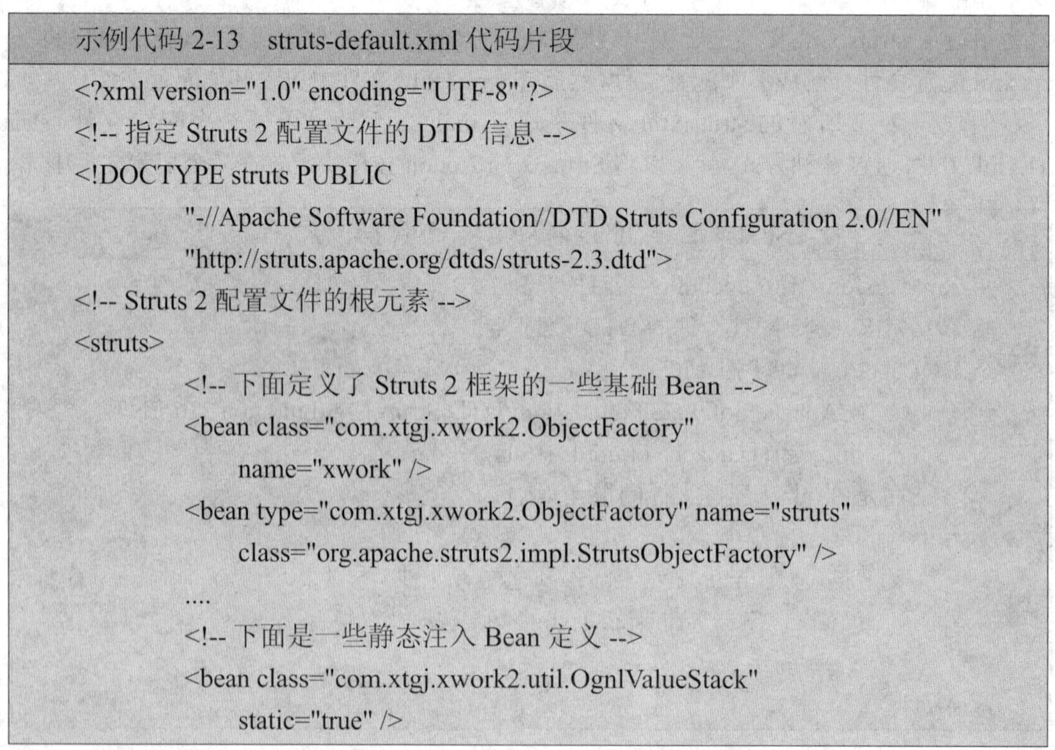

示例代码 2-13 struts-default.xml 代码片段

```
<?xml version="1.0" encoding="UTF-8" ?>
<!-- 指定 Struts 2 配置文件的 DTD 信息 -->
<!DOCTYPE struts PUBLIC
        "-//Apache Software Foundation//DTD Struts Configuration 2.0//EN"
        "http://struts.apache.org/dtds/struts-2.3.dtd">
<!-- Struts 2 配置文件的根元素 -->
<struts>
        <!-- 下面定义了 Struts 2 框架的一些基础 Bean -->
        <bean class="com.xtgj.xwork2.ObjectFactory"
            name="xwork" />
        <bean type="com.xtgj.xwork2.ObjectFactory" name="struts"
            class="org.apache.struts2.impl.StrutsObjectFactory" />
        ....
        <!-- 下面是一些静态注入 Bean 定义 -->
        <bean class="com.xtgj.xwork2.util.OgnlValueStack"
            static="true" />
```

```
<bean class="org.apache.struts2.dispatcher.Dispatcher"
    static="true" />
...
<!-- 下面定义 Struts 2 的默认包空间 -->
<package name="struts-default">
        <!-- 定义 Struts 2 内建支持的结果类型 -->
        <result-types>
            <!-- 定义 Action 链 Result 类型 -->
            <result-type name="chain"
                class="com.xtgj.xwork2.ActionChainResult" />
            <!-- 定义 Dispatcher 的 Result 类型，并设置 default="true",
                    指定该结果 Result 是默认的 Result 类型 -->
            <result-type name="dispatcher"
                class="org.apache.struts2.dispatcher.Servlet
                                                    DispatcherResult"
                default="true" />
            <!-- 定义 FreeMarker 的 Result 类型 -->
            <result-type name="freemarker"
                class="org.apache.struts2.views.freemarker.

                                                    FreemarkerResult" />
            ...
        </result-types>
        <!-- 定义 Struts 2 内建的拦截器 -->
        <interceptors>
    <interceptor name="alias"                class="com.xtgj.xwork2.inter-
ceptor.AliasInterceptor" />
        <interceptor name="autowiring"
    class="com.xtgj.xwork2.spring.interceptor.
  ActionAutowiringInterceptor" />
            ...
            <!-- 定义基本拦截器栈 -->
            <interceptor-stack name="basicStack">
                <interceptor-ref name="exception" />
                <interceptor-ref name="servlet-config" />
                <interceptor-ref name="prepare" />
                <interceptor-ref name="checkbox" />
                <interceptor-ref name="params" />
```

```
                <interceptor-ref name="conversionError" />
            </interceptor-stack>
            <!-- 还有系列拦截器栈 -->
            ...
        </interceptors>
        <!-- 定义默认的拦截器栈引用 -->
        <default-interceptor-ref name="defaultStack" />
    </package>
</struts>
```

以上代码并未全部列出 struts-default.xml 文件，只是列出了每个元素的代表。上述配置文件中定义了一个名字为 struts-default 的包空间，该包空间里定义了 Struts 2 内建的 result 类型，还定义了 Struts 2 内建的系列拦截器，以及由不同拦截器组成的拦截器栈，文件的最后还定义了默认的拦截器引用。

这个 struts-default.xml 文件是 Struts 2 框架的默认配置文件，Struts 2 框架每次都会自动加载该文件。查看前面使用的 struts.xml 文件，看到我们自己定义的 package 元素有如示例代码 2-14 所示的代码片段。

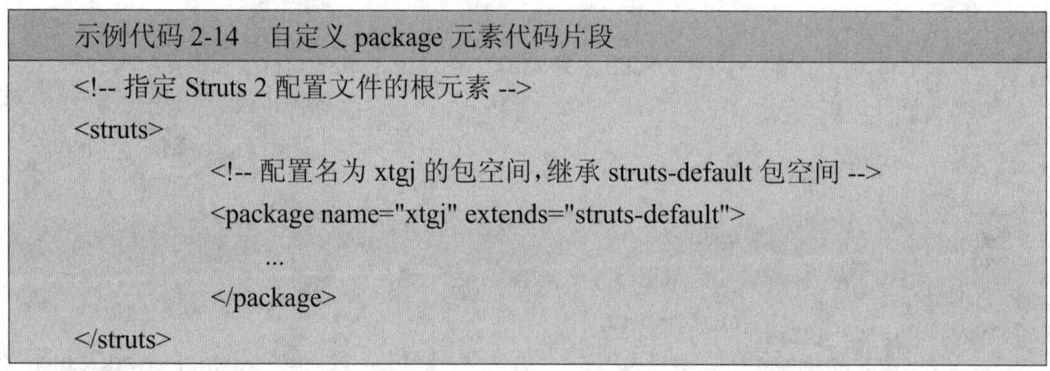

示例代码 2-14　自定义 package 元素代码片段

```
<!-- 指定 Struts 2 配置文件的根元素 -->
<struts>
        <!-- 配置名为 xtgj 的包空间，继承 struts-default 包空间 -->
        <package name="xtgj" extends="struts-default">
            ...
        </package>
</struts>
```

在上述配置文件中，名为 xtgj 的包空间，继承了名为 struts-default 的包空间，struts-default 包空间定义在 struts-default.xml 文件中。可见，Struts 2 框架默认会加载 struts-default.xml 文件。

2.3 小结

- ✓ 通过 MyEclipse 工具来开发 Struts 2 应用。
- ✓ Struts 2 的配置文件。
- ✓ 通过 web.xml 文件加载 Struts 2 框架。
- ✓ 一个简单的登录实现。

2.4　英语角

POJO　　普通的 Java 对象
default　默认
tag　　　标签

2.5　作业

根据本章内容，使用 Struts 2 框架完成用户登录。

2.6　思考题

思考 Struts 2 与 Struts 1 的区别。

2.7　学员回顾内容

1. 为 Web 应用增加 Struts 2 支持。
2.Struts 2 框架的流程。

第 3 章　深入了解 Struts 2

学习目标

✧ 熟悉 struts-default.xml 配置文件。
✧ 熟悉 struts.xml 配置文件。
✧ 掌握 Struts 2 的多模块设计。
✧ 掌握 Struts 2 的通配符方式定义 action。
✧ 掌握 Struts 2 的请求参数。
✧ 熟悉 struts.properties 配置文件。

课前准备

✧ 熟悉 Struts 1 的基本结构。
✧ 熟悉 Struts 2 的运行原理。

本章简介

本章主要介绍 Struts 2 的配置文件，学习配置文件的使用方法，了解 Struts 2 的运行流程以及它的基本原理。

3.1　Struts 2 中的配置文件

每学习一个框架，我们都免不了要学习一些关于这个框架的配置文件，Struts 2 也不例外，下文介绍 Struts 2 中几个主要的配置文件。Struts 2 的配置文件是以 XML 的形式出现的，不过它的 XML 的语义比较简单。Struts 2 中涉及的几个配置文件主要包括：

（1）struts-default.xml，这个文件是 Struts 2 框架默认加载的配置文件，它定义 Struts 2 的一些核心的 Bean 和拦截器。

（2）struts.xml，该文件也是 Struts 2 框架自动加载的文件，在这个文件中可以定义一些自己的 action、interceptor、package 等，该文件的 package 通常继承 struts-default 包。

（3）struts.properties 文件，这个文件是 Struts 2 框架的全局属性文件，也是自动加载的文件，该文件包含了一系列的 key-value 对。该文件完全可以配置在 struts.xml 文件中，使用 constant 元素实现。

接下来，我们将为大家简单介绍 Struts 2 中几个主要配置文件的用法及配置项的含义。

3.1.1 struts-default.xml

我们先来看一段 struts-default.xml 文件的代码，如示例代码 3-1 所示。

示例代码 3-1　struts-default.xml 配置文件的示例

```
<?xml version="1.0" encoding="UTF-8" ?>
<!DOCTYPE struts PUBLIC
"-//Apache Software Foundation//DTD Struts Configuration 2.3//EN"
"http://struts.apache.org/dtds/struts-2.3.dtd">
<struts>
<!--struts 2 中工厂 bean 的定义 -->
<bean class="com.opensymphony.xwork2.ObjectFactory" name="xwork" />
……
<!-- 类型检测 bean 的定义 -->
<bean type="com.opensymphony.xwork2.util.ObjectTypeDeterminer" name="tiger"
class="com.opensymphony.xwork2.util.GenericsObjectTypeDeterminer"/>
……
<!-- 文件上传 bean 的定义 -->
<bean type="org.apache.struts2.dispatcher.mapper.ActionMapper" name="struts"
class="org.apache.struts2.dispatcher.mapper.DefaultActionMapper" />
……
<!-- 标签库 bean 的定义 -->
<bean type="org.apache.struts2.views.TagLibrary" name="s" class="org.apache.struts2.
views.DefaultTagLibrary" />
<!-- 一些常用视图 bean 的定义 -->
<bean class="org.apache.struts2.views.freemarker.FreemarkerManager" name="struts"
optional="true"/>
……
<!-- 类型转换 bean 的定义 -->
<bean type="com.opensymphony.xwork2.util.XWorkConverter" name="xwork1"
class="com.opensymphony.xwork2.util.XWorkConverter" />
……
<!-- Struts 2 中一些可以静态注入的 bean, 也就是不需要实例化的 -->
<bean class="com.opensymphony.xwork2.ObjectFactory" static="true" />
……
<!-- 定义 Struts2 默认包 -->
<package name="struts-default" abstract="true">
```

```
<!-- 结果类型的种类 -->
<result-types>
<result-type name="chain" class="com.opensymphony.xwork2.ActionChainResult"/>
……
<result-type name="plaintext" class="org.apache.struts2.dispatcher.PlainTextResult" />
</result-types>

<!--struts2 中拦截器的定义 -->
<interceptors>
……
<!-- 负责类型转换的拦截器 -->
<interceptor name="conversionError" class="org.apache.struts2.interceptor.StrutsConversionErrorInterceptor"/>
……
<!-- 异常处理 -->
<interceptor name="exception" class="com.opensymphony.xwork2.interceptor.ExceptionMappingInterceptor"/>
<!-- 文件上传,解析表单域的内容 -->
<interceptor name="fileUpload" class="org.apache.struts2.interceptor.FileUploadInterceptor"/>
<!-- 支持国际化 -->
<interceptor name="i18n" class="com.opensymphony.xwork2.interceptor.I18nInterceptor"/>
<!-- 日志记录 -->
<interceptor name="logger" class="com.opensymphony.xwork2.interceptor.LoggingInterceptor"/>
……
<!-- 防止表单重复提交 -->
<interceptor name="token" class="org.apache.struts2.interceptor.TokenInterceptor"/>
……
<!-- 一个基本的拦截器栈 -->
<interceptor-stack name="basicStack">
<interceptor-ref name="exception"/>

<interceptor-ref name="servletConfig"/>
<interceptor-ref name="prepare"/>

<interceptor-ref name="checkbox"/>
```

```
<interceptor-ref name="params"/>
<interceptor-ref name="conversionError"/>
</interceptor-stack>

<!-- 简单的 validation 和 webflow 栈 -->
<interceptor-stack name="validationWorkflowStack">
<interceptor-ref name="basicStack"/>
<interceptor-ref name="validation"/>
<interceptor-ref name="workflow"/>
</interceptor-stack>

<!-- 文件上传的拦截器栈 -->
<interceptor-stack name="fileUploadStack">
<interceptor-ref name="fileUpload"/>
<interceptor-ref name="basicStack"/>
</interceptor-stack>

<!-- model-driven 栈  -->
<interceptor-stack name="modelDrivenStack">
<interceptor-ref name="modelDriven"/>
<interceptor-ref name="basicStack"/>
</interceptor-stack>

<!-- action 链的拦截器栈 -->
<interceptor-stack name="chainStack">
<interceptor-ref name="chain"/>
<interceptor-ref name="basicStack"/>
</interceptor-stack>

<!--  i18n 拦截器栈 -->
<interceptor-stack name="i18nStack">
<interceptor-ref name="i18n"/>
<interceptor-ref name="basicStack"/>
</interceptor-stack>
<!-- 结合 preparable 和 ModenDriven 拦截器 -->
<interceptor-stack name="paramsPrepareParamsStack">
<interceptor-ref name="exception"/>
……
```

```
<interceptor-ref name="workflow">
<param name="excludeMethods">input,back,cancel</param>
</interceptor-ref>
</interceptor-stack>

<!-- 定义默认的拦截器栈 -->
<interceptor-stack name="defaultStack">
<interceptor-ref name="exception"/>
……
<interceptor-ref name="execAndWait">
<param name="excludeMethods">input,back,cancel</param>
</interceptor-ref>
</interceptor-stack>

<interceptor name="external-ref" class="com.opensymphony.xwork2.interceptor.ExternalReferencesInterceptor"/>
……
<interceptor name="token-session" class="org.apache.struts2.interceptor.TokenSessionStoreInterceptor"/>

</interceptors>
<!-- 定义默认拦截器为 "defaultStack"-->
<default-interceptor-ref name="defaultStack"/>
</package>

</struts>
```

在这个配置文件中可以看到，Struts 2 的 XML 自身所支持的节点和子节点并不是很多，它定义 Struts 2 一些核心的 Bean 和拦截器。大致来说，这些节点可以分成基本配置定义和 Runtime 配置定义。

（1）基本配置定义

基本配置定义，主要是针对在 Struts 2 内部所使用的各种元素的声明。这些声明往往规定了 Struts 2 内部的一些行为特征。

例如，配置文件中的 <bean> 节点，被用于定义 Struts 2 中所使用的接口和实现类，通过 Struts 2 内部实现的控制反转，就可以在不同的实现类之间进行切换。再例如，配置文件中的 <result-type> 节点和 <int-erceptor> 节点，它们用于定义 Struts 2 中所支持的所有的 result 类型和拦截器，这些定义和声明将在 Runtime 的配置定义中被引用。

之所以把配置文件中的这些节点单独列出来，作为一个种类，是因为这些节点是不可省略的，也是无法简化的。所以，如果试图在 Struts 2 中简化配置，就需要认真研究 Runtime 配置定义，而这些基本配置定义是 Runtime 配置定义的基础。

（2）Runtime 配置定义

Runtime 配置定义，主要指的是 Struts 2 运行过程中，具体的某个 action 的行为的指定。这些指定主要通过 struts.xml 文件中的 <package> 节点中的 <action> 节点来完成。

3.1.2 struts.xml

我们再来看一段 struts.xml 文件的代码，如示例代码 3-2 所示。

示例代码 3-2　struts.xml 配置文件的示例

```
<?xml version="1.0" encoding="GBK"?>
<!DOCTYPE struts PUBLIC
"-//Apache Software Foundation//DTD Struts Configuration 2.3//EN"
"http://struts.apache.org/dtds/struts-2.3.dtd">
<!-- struts 是 Struts 2 配置文件的根元素 -->
<struts>
<!-- 下面元素用于配置常量，可以出现 0 次，也可以无限多次 -->
<constant name="" value="" />
<!-- 下面元素用于配置其他模块的配置文件，可以出现 0 次，也可以无限多次 -->
<include file="" />
<!-- package 元素是 Struts 配置文件的核心，该元素可以出现 0 次，或者无限多次
-->
<package name=" " extends="" namespace="" abstract="" >
<!-- 该元素可以出现，也可以不出现，最多出现一次 -->
<result-types>
<!-- 该元素必须出现，可以出现无限多次 -->
<result-type name="" class="" default="true|false">
<!-- 下面元素可以出现 0 次，也可以无限多次 -->
<param name=" "> </param>
</result-type>
</result-types>
<!-- 该元素可以出现，也可以不出现，最多出现一次 -->
<interceptors>
<!-- 该元素的 interceptor 元素和 interceptor-stack 至少出现其中之一，
也可以二者都出现 -->
<!-- 下面元素可以出现 0 次，也可以无限多次 -->
<interceptor name="" class="">
<!-- 下面元素可以出现 0 次，也可以无限多次 -->
<param name=" 参数名 "> 参数值 </param>
```

```
</interceptor>
<!-- 下面元素可以出现 0 次，也可以无限多次 -->
<interceptor-stack name="">
<!-- 该元素必须出现，可以出现无限多次 -->
<interceptor-ref name="">
<!-- 下面元素可以出现 0 次，也可以无限多次 -->
<param name=" 参数名 "> 参数值 </param>
</interceptor-ref>
</interceptor-stack>
</interceptors>
<!-- 下面元素可以出现 0 次，也可以无限多次 -->
<default-interceptor-ref name="">
<!-- 下面元素可以出现 0 次，也可以无限多次 -->
<param name=" 参数名 "> 参数值 </param>
</default-interceptor-ref>
<!-- 下面元素可以出现 0 次，也可以无限多次 -->
<default-action-ref name="">
<!-- 下面元素可以出现 0 次，也可以无限多次 -->
<param name=" 参数名 "> 参数值 </param>
</default-action-ref>
<!-- 下面元素可以出现 0 次，也可以无限多次 -->
<global-results>
<!-- 该元素必须出现，可以出现无限多次 -->
<result name="" type="">
<!-- 该字符串内容可以出现 0 次或多次 -->
映射资源
<!-- 下面元素可以出现 0 次，也可以无限多次 -->
<param name=" 参数名 "> 参数值 </param>
</result>
</global-results>
<!-- 下面元素可以出现 0 次，也可以无限多次 -->
<global-exception-mappings>
<!-- 该元素必须出现，可以出现无限多次 -->
<exception-mapping name="" exception="" result="">
异常处理资源
<!-- 下面元素可以出现 0 次，也可以无限多次 -->
<param name=" 参数名 "> 参数值 </param>
</exception-mapping>
```

```
</global-exception-mappings>
<action name="" class="" method="" converter="">
<!-- 下面元素可以出现 0 次，也可以无限多次 -->
<param name=" 参数名 "> 参数值 </param>
<!-- 下面元素可以出现 0 次，也可以无限多次 -->
<result name="" type="">
映射资源
<!-- 下面元素可以出现 0 次，也可以无限多次 -->
<param name=" 参数名 "> 参数值 </param>
</result>
<!-- 下面元素可以出现 0 次，也可以无限多次 -->
<interceptor-ref name="">
<!-- 下面元素可以出现 0 次，也可以无限多次 -->
<param name=" 参数名 "> 参数值 </param>
</interceptor-ref>
<!-- 下面元素可以出现 0 次，也可以无限多次 -->
<exception-mapping name="" exception="" result="">
异常处理资源
<!-- 下面元素可以出现 0 次，也可以无限多次 -->
<param name=" 参数名 "> 参数值 </param>
</exception-mapping>
</action>
</package>
<struts>
```

1. struts.xml 配置中的包

在 Struts 2 框架中使用包来管理 action，包的作用和 Java 中的类包是非常类似的，它主要用于管理一组业务功能相关的 action。在实际应用中，我们应该把一组业务功能相关的 action 放在同一个包下。

配置包时必须指定 name 属性，该 name 属性值可以任意取名，但必须唯一，它不对应 Java 的类包，如果其他包要继承该包，必须通过该属性进行引用。包的 namespace 属性用于定义该包的命名空间，命名空间作为访问该包下 action 的路径的一部分，在 Struts 1 中，通过 <action path="/test/helloworld"> 节点的 path 属性指定访问该 action 的 URL 路径。在 Struts 2 中，情况则不同，访问 Struts 2 中 action 的 URL 路径由两部分组成："包的命名空间名称 +action 的名称"，例如访问下述案例中 HelloWorldAction 的 URL 路径为"/test/helloworld"（注意：完整路径为：http://localhost: 端口 / 内容路径 /test/helloworld）。另外我们也可以加上 .action 后缀访问此 action（/test/helloworld.action）。 namespace 属性可以不配置，对本例而言，如果不指定该属性，默认的命名空间为" "（空字符串）。

示例代码 3-3　struts.xml 配置文件中的包

```
<package name="xtgj" namespace="/test" extends="struts-default">
    <action name="helloworld" class="com.xtgj.action.HelloWorldAction" method="
execute" >
            <result name="success">/WEB-INF/page/hello.jsp</result>
    </action>
</package>
```

通常每个包都应该继承 struts-default 包，因为 Struts 2 很多核心的功能都是通过拦截器来实现，如从请求中把请求参数封装到 action，文件上传和数据验证等都是通过拦截器实现的。struts-default 定义了这些拦截器和 result 类型。可以这么说：当包继承了 struts-default 才能使用 Struts 2 提供的核心功能。Struts-default 包是在 struts 2-core-2.x.x.jar 文件中的 struts-default.xml 中定义的。struts-default.xml 也是 Struts 2 默认配置文件。Struts 2 每次都会自动加载 struts-default.xml 文件。

包还可以通过 abstract="true"定义为抽象包，抽象包中不能包含 action。

2. Action 名称的搜索顺序

（1）获得请求路径的 URL，例如 URL 为：http://server/struts2/path1/path2/path3/test.action。

（2）寻找 namespace 为 /path1/path2/path3 的 package，如果不存在这个 package 则执行步骤 3；如果存在这个 package，则在这个 package 中寻找名字为 test 的 action，当在该 package 下寻找不到 action 时就会直接到默认 namespace 的 package 中寻找 action（默认的命名空间为空字符串""），如果在默认 namespace 的 package 中仍寻找不到该 action，页面提示找不到 action。

（3）寻找 namespace 为 /path1/path2 的 package，如果不存在这个 package，则转至步骤 4；如果存在这个 package，则在这个 package 中寻找名字为 test 的 action，当在该 package 中寻找不到 action 时就会直接到默认 namespace 的 package 中寻找名为 test 的 action ，在默认 namespace 的 package 中仍寻找不到该 action，页面提示找不到 action。

（4）寻找 namespace 为 /path1 的 package，如果不存在这个 package 则执行步骤 5；如果存在这个 package，则在这个 package 中寻找名字为 test 的 action，当在该 package 中寻找不到 action 时就会直接到默认 namespace 的 package 中寻找名字为 test 的 action ，在默认 namespace 的 package 中仍寻找不到该 action，页面提示找不到 action；

（5）寻找 namespace 为 / 的 package，如果存在这个 package，则在这个 package 中寻找名字为 test 的 action，当在 package 中寻找不到 action 或者不存在这个 package 时，都会去默认 namespace 的 package 中寻找 action，如果仍找不到，页面提示找不到 action。

3. Action 配置中的各项默认值

（1）如果没有为 action 指定 class，默认是 ActionSupport。

（2）如果没有为 action 指定 method，默认执行 action 中的 execute() 方法。

（3）如果没有指定 result 的 name 属性，默认值为 success。

4. Action 中 result 的各种转发类型

result 配置类似于 Struts 1 中的 forward，但 Struts 2 中提供了多种结果类型，常用的类型有：dispatcher（默认值）、redirect、redirectAction、plainText。例如 dispatcher 结果类型的配置如示例代码 3-4 所示。

示例代码 3-4　Action 转发类型

```
<action name="helloworld" class="com.xtgj.action.HelloWorldAction">
    <result name="success">/WEB-INF/page/hello.jsp</result>
</action>
```

在 result 中还可以使用 ${ 属性名 } 表达式访问 action 中的属性，表达式里的属性名对应 action 中的属性。例如 redirect 结果类型的配置如下：

```
<result type="redirect">/view.jsp?id=${id}</result>
```

以下是 redirectAction 结果类型的例子，如果重定向的 action 中同一个命名空间下：

```
<result type="redirectAction">helloworld</result>
```

如果重定向的 action 在别的命名空间下，则按照以下方式配置：

```
<result type="redirectAction">
    <param name="actionName">helloworld</param>
    <param name="namespace">/test</param>
</result>
```

plainText 结果类型用于显示原始文件内容，例如当我们需要原样显示 jsp 文件源代码的时候，我们可以使用此类型。

```
<result name="source" type="plainText ">
    <param name="location">/xxx.jsp</param>
    <param name="charSet">UTF-8</param><!-- 指定读取文件的编码 -->
</result>
```

5. 全局 result 配置

当多个 action 中都使用到了相同视图，这时我们应该把 result 定义为全局视图。Struts 1 中提供了全局 forward，Struts 2 中也提供了相似功能，例如：

```
<package ……>
        <global-results>
            <result name="message">/message.jsp</result>
        </global-results>
</package>
```

6. 为 action 的属性注入值

Struts 2 为 action 中的属性提供了依赖注入功能，在 Struts 2 的配置文件中，我们可以很方便地为 action 中的属性注入值。注意：属性必须提供 setter 方法。HelloWorldAction 类的定义如示例代码 3-5 所示。

示例代码 3-5 action 中的属性注入

```
public class HelloWorldAction{
        private String savePath;
        public String getSavePath() {
                return savePath;
        }
        public void setSavePath(String savePath) {
                this.savePath = savePath;
        }
}
```

可以通过 <param> 节点为 action 的 savePath 属性注入“/images”，例如：

```
<package name="xtgj" namespace="/test" extends="struts-default">
    <action name="helloworld" class="com.xtgj.action.HelloWorldAction" >
        <param name="savePath">/images</param>
        <result name="success">/WEB-INF/page/hello.jsp</result>
    </action>
</package>
```

7. 指定需要 Struts 2 处理的请求后缀

前面我们都是默认使用 .action 后缀访问 action。其实默认后缀是可以通过常量“struts.action.extension”进行修改的，例如我们可以配置 Struts 2 只处理以 .do 为后缀的请求路径。

```
<struts>
        <constant name="struts.action.extension" value="do"/>
</struts>
```

如果用户需要指定多个请求后缀，则多个后缀之间以英文逗号(,)隔开。如：

```
<constant name="struts.action.extension" value="do,go"/>
```

8. Struts 2 **中常用的常量介绍**

```
<!-- 指定默认编码集 , 作用于 HttpServletRequest 的 setCharacterEncoding 方法 和 freemarker 、velocity 的输出 -->
<constant name="struts.i18n.encoding" value="UTF-8"/>
<!-- 该属性指定需要 Struts 2 处理的请求后缀，该属性的默认值是 action，即所有匹配 *.action 的请求都由 Struts2 处理。如果用户需要指定多个请求后缀，则多个后缀之间以英文逗号(,)隔开。 -->
<constant name="struts.action.extension" value="do"/>
<!-- 设置浏览器是否缓存静态内容 , 默认值为 true( 生产环境下使用 ), 开发阶段最好关闭 -->
<constant name="struts.serve.static.browserCache" value="false"/>
<!-- 当 struts 的配置文件修改后 , 系统是否自动重新加载该文件 , 默认值为 false( 生产环境下使用 ), 开发阶段最好打开 -->
<constant name="struts.configuration.xml.reload" value="true"/>
<!-- 开发模式下使用 , 这样可以打印出更详细的错误信息 -->
<constant name="struts.devMode" value="true" />
<!-- 默认的视图主题 -->
<constant name="struts.ui.theme" value="simple" />
<!– 与 spring 集成时，指定由 spring 负责 action 对象的创建 -->
<constant name="struts.objectFactory" value="spring" />
<!– 该属性设置 Struts 2 是否支持动态方法调用，该属性的默认值是 true。如果需要关闭动态方法调用，则可设置该属性为 false。 -->
<constant name="struts.enable.DynamicMethodInvocation" value="false"/>
 <!-- 上传文件的大小限制 -->
<constant name="struts.multipart.maxSize" value="10701096"/>
```

9. **指定多个** Struts **配置文件**

在大部分应用里，随着应用规模的增加，系统中 action 的数量也会大量增加，导致 struts.xml 配置文件变得非常臃肿。为了避免 struts.xml 文件过于庞大、臃肿，提高 struts.xml 文件的可读性，我们可以将一个 struts.xml 配置文件分解成多个配置文件，然后在 struts.xml 文件中包含其他配置文件。以下的 struts.xml 通过 <include> 元素指定多个配置文件，如示例代码 3-6 所示。

示例代码 3-6　include 元素的使用

```
<?xml version="1.0" encoding="UTF-8"?>
<!DOCTYPE struts PUBLIC
    "-//Apache Software Foundation//DTD Struts Configuration 2.3//EN"
    "http://struts.apache.org/dtds/struts-2.3.dtd">
```

```
<struts>
    <include file="struts-user.xml"/>
    <include file="struts-order.xml"/>
</struts>
```

其中，struts-user.xml 和 struts-order.xml 都是类似于 struts.xml 的配置文件，它们的文档结构需符合 struts.xml 的 DTD，我们可以认为它们分别对应了 user 模块和 order 模块，通过这种方式，我们就可以将 Struts 2 的 action 按模块添加在多个配置文件中。

10. Struts 2 中的动态方法调用

默认情况下，如果没有为 action 指定 method，则直接执行 action 中的 execute() 方法，但是，action 中存在多个方法时，如何通过一个 <action> 节点的配置匹配不同的方法呢？我们可以使用“!+ 方法名”的形式调用指定方法。action 类的代码如示例代码 3-7 所示。

示例代码 3-7　动态方法调用

```
public class HelloWorldAction{
        private String message;
        public String execute() throws Exception{
            this.message = " 第一个方法 ";
            return "success";
        }

        public String other() throws Exception{
            this.message = " 第二个方法 ";
            return "success";
        }
}
```

假设访问上面 action 的 URL 路径为：“/struts/test/helloworld.action”，要访问 action 的 other() 方法，我们可以这样调用：

```
/struts/test/helloworld!other.action
```

通常不建议读者使用动态方法调用。如果不使用动态方法调用，我们可以通过常量 struts. enable. DynamicMethodInvocation 关闭动态方法调用。

```
<constant name="struts.enable.DynamicMethodInvocation" value="false"/>
```

11. 使用通配符定义 action

示例代码 3-8　使用通配符

```
<package name="xtgj" namespace="/test" extends="struts-default">
    <action  name="helloworld_*"  class="com.xtgj.action.HelloWorldAction"  meth-
od="{1}">
            <result name="success">/WEB-INF/page/hello.jsp</result>
    </action>
</package>
```

<action> 节点的 name 属性被赋值为“helloworld_*”，其中的下划线（_）并不是必需的，只是一种命名的习惯。因为 helloworld_* 中只有一个通配符 *，所以 method 属性赋值为 "{1}"，此时要访问 action 中的 other() 方法，可以通过这样的 URL 访问：“/test/helloworld_other.action”。

当然，name 也可以是 helloworld_*_*，则 method="{1}_{2}"，class 和 result 中也可以使用通配符的值，如：

```
“class="com.xtgj.action.{1}HelloWorldAction"”
“<result name="success">/WEB-INF/page/{1}hello.jsp</result>”
```

12. 接收请求参数

（1）采用基本类型接收请求参数

在 action 类中定义与请求参数同名的属性，Struts 2 便能自动接收请求参数并赋予给同名属性。假设有如下请求路径：http://localhost:8080/test/view.action?id=78，如示例代码 3-9 所示：

示例代码 3-9　接受请求参数

```
public class ProductAction {
        private Integer id;
        //struts2 通过反射技术调用与请求参数同名的属性的 setter 方法来获取请求
参数值
        public void setId(Integer id) {
                this.id = id;
        }
        public Integer getId() {return id;}
}
```

（2）采用复合类型接收请求参数

在 action 类中定义对象成员，该对象封装了自己的属性，请求参数采取“对象名 . 属性”的形式，Struts 2 便能自动接收复合类型的请求参数，并初始化属性值到对应对象，例如有如下请求路径：http://localhost:8080/test/view.action?product.id=78 ，代码如示例代码 3-10 所示。

示例代码 3-10　接受复合类型请求参数

```
public class ProductAction {
    private Product product;
    public void setProduct(Product product) {  this.product = product;  }
    public Product getProduct() {return product;}
}
```

Struts 2 首先通过反射技术调用 Product 的默认构造器创建 product 对象，然后再通过反射技术调用 Product 中与请求参数同名的属性的 setter 方法来获取请求参数值。

3.1.3　struts.properties

我们来看一段 struts.properties 文件的代码片段，由于该文件可以用 struts.xml 替代，所以不赘述。

示例代码 3-11　属性文件示例

```
### 指定加载 struts2 配置文件管理器，默认为 org.apache.struts2.config.DefaultCon-
figuration
### 开发者可以自定义配置文件管理器，该类要实现 Configuration 接口，可以自动
加载 struts2 配置文件。
# struts.configuration=org.apache.struts2.config.DefaultConfiguration
### 设置默认的 locale 和字符编码
# struts.locale=en_US
struts.i18n.encoding=UTF-8
### 指定 struts 的工厂类
# struts.objectFactory = spring
### 指定 spring 框架的装配模式
### 装配方式有 : name, type, auto, and constructor (name 是默认装配模式 )
struts.objectFactory.spring.autoWire = name
### 该属性指定整合 spring 时，是否对 bean 进行缓存，值为 true or false, 默认为
true.
struts.objectFactory.spring.useClassCache = true
### 指定类型检查
#struts.objectTypeDeterminer = tiger
#struts.objectTypeDeterminer = notiger
### 该属性指定处理 MIME-type multipart/form-data，文件上传
# struts.multipart.parser=cos
# struts.multipart.parser=pell
struts.multipart.parser=jakarta
```

```
# 指定上传文件时的临时目录，默认使用 javax.servlet.context.tempdir
struts.multipart.saveDir=
struts.multipart.maxSize=2097152
### 加载自定义属性文件 ( 不要改写 struts.properties!)
# struts.custom.properties=application,org/apache/struts2/extension/custom
### 指定请求 url 与 action 映射器，默认为 org.apache.struts2.dispatcher.mapper.DefaultActionMapper
#struts.mapper.class=org.apache.struts2.dispatcher.mapper.DefaultActionMapper
### 指定 action 的后缀，默认为 action
struts.action.extension=action
### 被 FilterDispatcher 使用
### 如果为 true 则通过 jar 文件提供静态内容服务
### 如果为 false 则静态内容必须位于 <context_path>/struts
struts.serve.static=true
### 被 FilterDispatcher 使用
### 指定浏览器是否缓存静态内容，测试阶段设置为 false，发布阶段设置为 true.
struts.serve.static.browserCache=true
### 设置是否支持动态方法调用，true 为支持，false 不支持 .
struts.enable.DynamicMethodInvocation = true
### 设置是否可以在 action 中使用斜线，默认为 false 不可以，想使用需设置为 true.
struts.enable.SlashesInActionNames = false
### 是否允许使用表达式语法，默认为 true.
struts.tag.altSyntax=true
### 设置当 struts.xml 文件改动时，是否重新加载 .
### - struts.configuration.xml.reload = true
### 设置 struts 是否为开发模式，默认为 false, 测试阶段一般设为 true.
struts.devMode = false
### 设置是否每次请求，都重新加载资源文件，默认值为 false.
struts.i18n.reload=false
### 标准的 UI 主题
### 默认的 UI 主题为 xhtml, 可以为 simple,xhtml 或 ajax.
struts.ui.theme=xhtml
### 模板目录
struts.ui.templateDir=template
# 设置模板类型，可以为 ftl, vm, or jsp.

struts.ui.templateSuffix=ftl
### 定位 velocity.properties 文件，默认 velocity.properties.
```

```
struts.velocity.configfile = velocity.properties
### 设置 velocity 的 context
struts.velocity.contexts =
### 定位 toolbox
struts.velocity.toolboxlocation=
### 指定 web 应用的端口
struts.url.http.port = 80
### 指定加密端口
struts.url.https.port = 443
### 设置生成 url 时，是否包含参数，值可以为 : none, get or all.
struts.url.includeParams = get
### 设置要加载的国际化资源文件，以逗号分隔 .
# struts.custom.i18n.resources=testmessages,testmessages2
### 对于一些 web 应用服务器不能处理 HttpServletRequest.getParameterMap()
### 像 WebLogic, Orion, and OC4J 等，须设置成 true, 默认为 false.
struts.dispatcher.parametersWorkaround = false
### 指定 freemarker 管理器
#struts.freemarker.manager.classname=org.apache.struts2.views.freemarker.FreemarkerManager
### 设置是否对 freemarker 的模板设置缓存
### 效果相当于把 template 拷贝到 WEB_APP/templates.
struts.freemarker.templatesCache=false
### 通常不需要修改此属性
struts.freemarker.wrapper.altMap=true
### 指定 xslt result 是否使用样式表缓存，开发阶段设为 true, 发布阶段设为 false.
struts.xslt.nocache=false
### 设置 struts 自动加载的文件列表
struts.configuration.files=struts-default.xml,struts-plugin.xml,struts.xml
### 设定是否一直在最后一个 slash 之前的任何位置选定 namespace
struts.mapper.alwaysSelectFullNamespace=false
```

在上机部分，我们实现了部门和商品两个模块的业务操作，请读者认真阅读并实践。

3.2 小结

- ✓ Struts 2 的配置文件是以 XML 的形式出现的，不过它的 XML 的语义比较简单。
- ✓ Struts 2 中涉及的几个配置文件主要包括：struts-default.xml、struts.xml、struts.properties。

✓ struts-default.xml 是 Struts 2 框架默认加载的配置文件，它定义 Struts 2 的一些核心的 bean 和拦截器。其中的节点可以分成基本配置定义和 Runtime 配置定义。

✓ struts.xml 也是 Struts 2 框架默认加载的文件，在这个文件中可以自定义一些 action、interceptor、package 等。

✓ struts.properties 文件是 Struts 2 框架的全局属性文件，也是自动加载的文件，该文件包含了一系列的 key-value 对。

3.3　英语角

properties	属性
runtime	运行
namespace	命名空间
extension	延时

3.4　作业

1. 简述 Struts 2 的三个主要配置文件的作用。
2. 根据本章内容，使用 Struts 2 框架完成用户登录、注册、修改个人资料的业务。

3.5　思考题

接收到的中文请求参数为乱码的问题在 Struts 2 中是否已经得到解决？是如何处理的？

3.6　学员回顾内容

1.Struts 2 中涉及的配置文件包括：struts-default.xml、struts.xml、struts.properties。
2.struts-default.xml 的作用。
3.struts.xml 的作用。
4.struts.properties 的作用。

第 4 章　Struts 2 转换器

学习目标

- ✧ 了解转换器的主要特性。
- ✧ 理解转换器的组成结构。
- ✧ 掌握转换器的使用。

课前准备

- ✧ Struts 2 转换器的组成。
- ✧ Struts 2 转换器的分类。

本章简介

本章主要介绍 Struts 2 转换器的主要特性，学习转换器的组成结构，了解转换器的使用方法以及它的分类。

4.1　转换器介绍

4.1.1　概述

开发 Web 应用程序与开发传统桌面应用程序不同，Web 应用程序实际上是分布在不同的主机（当然也可以是同一个主机，不过比较少见）上的两个进程之间互交。这种互交建立在 HTTP 之上，它们互相传递的都是字符串。换句话说，服务器接收到来自用户的数据只能是字符串或字符数组，而在服务器上的对象中，这些数据往往有多种不同的类型，如日期（Date）、整数（int）、浮点数（float）或自定义类型（UDT），如图 4-1 所示。因此，我们需要服务器端将字符串转换为适合的类型。

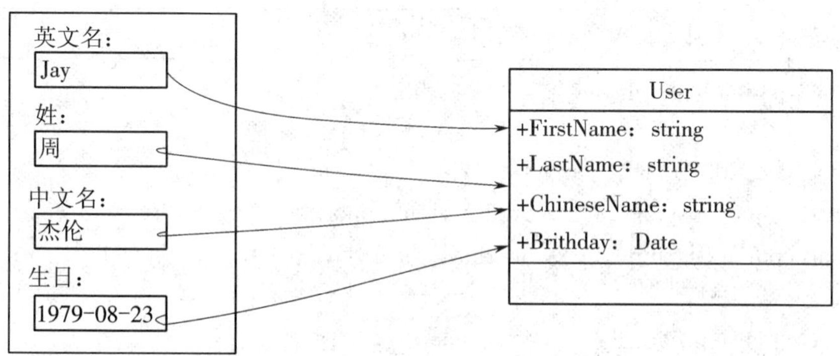

图 4-1　UI 与服务器对象关系

同样的问题也发生在使用 UI 展示服务器数据的情况。HTML 的 Form 控件不同于桌面应用程序可以表示对象，其值只能为字符串类型，所以我们需要通过某种方式将特定对象转换成字符串。要实现上述转换，Struts 2 中有位魔术师可以帮到你——Converter。有了它，就不用一遍又一遍地重复编写诸如此类代码。

```
Date birthday = DateFormat.getInstance(DateFormat.SHORT).parse(strDate);
<input type="text" value="<%= DateFormat.getInstance(DateFormat.SHORT).format(birthday) %>" />
```

4.1.2　实例介绍

现在请读者阅读下面的例子，逐步体会转换器的使用方式和它的作用。

首先，创建和配置默认的资源文件。在 MyEclipse 中创建工程配置开发和运行环境，在 src 文件夹中加入 globalMessages_en_US.properties 文件，内容如下：

```
HelloWorld=Hello World!
```

在 src 文件夹中加入 globalMessages_zh_CN.properties 文件，内容如下：

```
HelloWorld= 你好，世界！
```

其次，新建源代码文件夹下的 com.xtgj.struts2.chapter04.action 包创建 HelloWorld.java 文件，如示例代码 4-1 所示。

示例代码 4-1 HelloWorld.java 代码

```
package com.xtgj.struts2.chapter04.action;

import java.util.Locale;
import com.opensymphony.xwork2.ActionSupport;
import com.opensymphony.xwork2.util.LocalizedTextUtil;

public class HelloWorld {
    private String msg;
    private Locale loc = Locale.US;

    public String getMsg() {
        return msg;
    }

    public Locale getLoc() {
        return loc;
    }

    public void setLoc(Locale loc) {
        this.loc = loc;
    }

    public String execute() {
        // LocalizedTextUtil 是 Struts 2.0 中国际化的工具类，<s:text> 标志就是
通过调用它实现国际化的
        msg = LocalizedTextUtil.findDefaultText("HelloWorld", loc);
        return SUCCESS;
    }
}
```

然后，在源代码文件夹下的 struts.xml 加入如下代码注册 action，如示例代码 4-2 所示。

示例代码 4-2 struts.xml 代码

```
<?xml version="1.0" encoding="UTF-8" ?>
<!-- 指定 Struts 2 配置文件的 DTD 信息 -->
<!DOCTYPE struts PUBLIC
    "-//Apache Software Foundation//DTD Struts Configuration 2.0//EN"
    "http://struts.apache.org/dtds/struts-2.0.dtd">
<!-- struts 是 Struts 2 配置文件的根元素 -->
<struts>
    <constant name="struts.custom.i18n.resources"
        value="globalMessages" />
    <package name="ConverterDemo" extends="struts-default">
        <action name="HelloWorld"
            class="com.xtgj.struts2.chapter04.action.HelloWorld">
            <result>/HelloWorld.jsp</result>
        </action>
    </package>
</struts>
```

新建 HelloWorld.jsp，如示例代码 4-3 所示。

示例代码 4-3 HelloWorld.jsp

```
<%@ page contentType="text/html; charset=UTF-8"%>
<%@ taglib uri="/struts-tags" prefix="s"%>
<html>
    <head>
        <title>Hello World</title>
    </head>
    <body>
        <s:form action="HelloWorld" theme="simple">
      Locale: <s:textfield name="loc" />   <s:submit />
        </s:form>
        <h2>
            <s:property value="msg" />
        </h2>
    </body>
</html>
```

在源代码文件夹的 com.xtgj.struts2.chapter04.action 包中新建 LocaleConverter.java 文件，如示例代码 4-4 所示。

示例代码 4-4　LocaleConverter.java 代码

```
package com.xtgj.struts2.chapter04.action;

import java.util.Locale;
import java.util.Map;

public class LocaleConverter extends ognl.DefaultTypeConverter {
	@Override
	public Object convertValue(Map context, Object value, Class toType) {
		if (toType == Locale.class) {
			String locale = ((String[]) value)[0];
			return new Locale(locale.substring(0, 2), locale.substring(3));
		} else if (toType == String.class) {
			Locale locale = (Locale) value;
			return locale.toString();
		}
		return null;
	}
}
```

接下来，在源代码文件夹下新建 xwork-conversion.properties，并在其中添加如下代码：

```
java.util.Locale = com.xtgj.struts2.chapter04.action.LocaleConverter
```

发布运行应用程序，在浏览器中键入 http://localhost:8080/ ConverterTest/HelloWorld.jsp，在文本框中录入“en_US”，输出页面如图 4-2 所示。

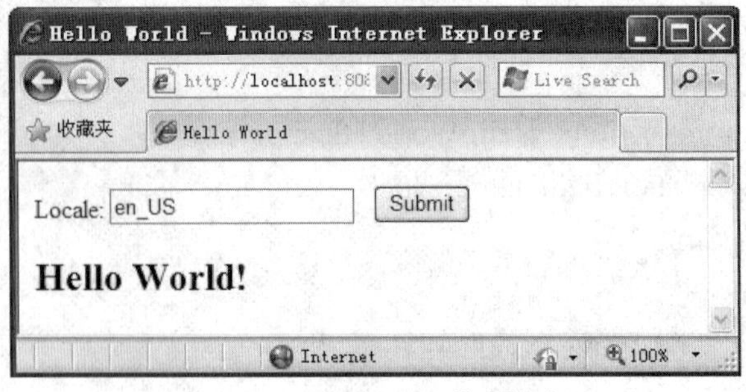

图 4-2　HelloWorld 英文输出

在 Locale 输入框中输入“zh_CN”，按“Submit”提交，出现如图 4-3 所示页面。

图 4-3　HelloWorld 中文输出

上述例子中，Locale 文本输入框对应是 action 中的类型为 java.util.Locale 的属性 loc，所以需要创建一个自定义转换器实现两者间的转换。所有的 Struts 2 中的转换器都必须实现 ognl.TypeConverter 接口。

为使操作简单，OGNL 包也提供了 ognl.DefaultTypeConverter 类去帮助用户实现转换器。在上述例子中，LocaleConverter 继承了 ognl.DefaultTypeConverter，重载了其方法原型为"public Object convertValue(Map context, Object value, Class toType)"的方法。下面简单地介绍一下函数的参数。

- Context：用于获取当前的 ActionContext；
- Value：需要转换的值；
- toType：需要转换成的目标类型。

想要实现转换器，我们需要通过配置告诉 Struts 2，可以通过以下两种方法做到：

- 对于配置全局的类型转换器，即上例的做法：在源代码文件夹下，新建一个名为"xwork-conversion.properties"的配置文件，并在文件中加入"待转换的类型的全名（包括包路径和类名）= 转换器类的全名"对。
- 对于应用于某个特定类的类型转换器，做法为在该类的包中添加一个格式为"类名 -conversion.properties"的配置文件，并在文件中加入"待转换的属性的名字 = 转换器类的全名"。对以上的例子也可以这样配置——在源代码文件夹的 com.xtgj.struts2.chapter04.action 包下新建名为"HelloWorld-conversion.properties"文件，并在其中加入：

```
loc= com.xtgj.struts2.chapter04.action.LocaleConverter
```

注意：在继承 DefaultTypeConverter 时，如果是要将 value 转换成其他非字符串类型时，要记住 value 是 String[] 类型，而不是 String 类型。它是通过 request.getParameterValues(String arg) 来获得的，所以不要试图将其强行转换为 String 类型。

4.1.3　已有的转换器

对于一些经常用到的转换器，如日期、整数或浮点数等类型，Struts 2 已经实现了。下面列出已经实现的转换器：

- 预定义类型，例如 int、boolean、double 等；
- 日期类型，使用当前区域（Locale）的短格式转换，即 DateFormat.getInstance(DateFor-

mat.SHORT)；

- 集合（Collection）类型，将 request.getParameterValues(String arg) 返回的字符串数据与 java.util.Collection 转换；
- 集合（Set）类型，与 List 的转换相似，去掉相同的值；
- 数组（Array）类型，将字符串数组的每一个元素转换成特定的类型，并组成一个数组。

对于已有的转换器，大家不必再去重新发明类型。Struts 2 在遇到这些类型时，会自动去调用相应的转换器。

4.2　批量封装对象

在使用 Struts 1.x 的时候，大家应该遇过这种情况，在一个页面里同时提交几个对象。例如，在发布产品的页面，同时发布几个产品。那对于程序员来说可谓是一个噩梦，幸运的是，在 Struts 2 中这种情况已经得到解决。下面就演示一下如何实现这个需求。

首先，在源代码文件夹下的 com.xtgj.struts2.chapter04.action 包中新建 Product.java 文件，如示例代码 4-5 所示。

示例代码 4-5　Product.java 代码

```
package com.xtgj.struts2.chapter04.action;
import java.util.Date;
public class Product {
    private String name;
    private double price;
    private Date dateOfProduction;
    public Date getDateOfProduction() {
        return dateOfProduction;
    }
    public void setDateOfProduction(Date dateOfProduction) {
        this.dateOfProduction = dateOfProduction;
    }
    public String getName() {
        return name;
    }
    public void setName(String name) {
        this.name = name;
```

```
    }
    public double getPrice() {
        return price;
    }
    public void setPrice(double price) {
        this.price = price;
    }
}
```

然后，在同上的包下添加 ProductConfirm.java 类，如示例代码 4-6 所示。

示例代码 4-6　ProductConfirm.java 代码

```
package com.xtgj.struts2.chapter04.action;

import java.util.List;
import com.opensymphony.xwork2.ActionSupport;

public class ProductConfirm extends ActionSupport {
    public List<Product> products;

    public List<Product> getProducts() {
        return products;
    }

    public void setProducts(List<Product> products) {
        this.products = products;
    }

    @Override
    public String execute() {
        for (Product p : products) {
            System.out.println(p.getName() + " | " + p.getPrice() + " | "
                    + p.getDateOfProduction());
        }
        return SUCCESS;
    }
}
```

接着，在同上的包中加入 ProductConfirm-conversion.properties，代码如下：

```
Element_products=com.xtgj.struts2.chapter04.action.Product
```

再在 struts.xml 文件中配置 ProductConfirm，如示例代码 4-7 所示。

示例代码 4-7　在 struts.xml 文件中配置 ProductConfirm

```
<action name="ProductConfirm"
            class="com.xtgj.struts2.chapter04.action.ProductConfirm">
            <result>/ShowProducts.jsp</result>
</action>
```

在 WEB-INF 文件夹下新建 AddProducts.jsp，如示例代码 4-8 所示。

示例代码 4-8　AddProducts.jsp 代码

```
<%@ page contentType="text/html; charset=UTF-8"%>
<%@taglib prefix="s" uri="/struts-tags"%>
<html>
    <head>
        <title> 添加商品 </title>
    </head>
    <body>
        <s:form action="ProductConfirm" theme="simple">
            <table>
                <tr style="background-color: powderblue; font-weight: bold;">
                    <td>
                        Product Name
                    </td>
                    <td>
                        Price
                    </td>
                    <td>
                        Date of production
                    </td>
                </tr>
                <s:iterator value="new int[3]" status="stat">
                    <tr>
                        <td>
                    <s:textfield name="%{'products['+#stat.index+'].name'}" />
                        </td>
```

```
                                    <td>
                              <s:textfield name="%{'products['+#stat.index+'].price'}" />
                                    </td>
                                    <td>
                                        <s:textfield
                               name="%{'products['+#stat.index+'].dateOfProduction'}" />
                                    </td>
                              </tr>
                          </s:iterator>
                          <tr>
                                <td colspan="3">
                                    <s:submit />
                                </td>
                          </tr>
                    </table>
                </s:form>
          </body>
    </html>
```

在同样的文件夹下创建 ShowProducts.jsp，内容如示例代码 4-9 所示。

示例代码 4-9　ShowProducts.jsp 代码

```
<%@ page contentType="text/html; charset=UTF-8"%>
<%@taglib prefix="s" uri="/struts-tags"%>
<html>
      <head>
            <title> 添加商品 </title>
      </head>
      <body>
            <table>
                  <tr style="background-color: powderblue; font-weight: bold;">
                        <td>
                              Product Name
                        </td>
                        <td>
                              Price
                        </td>
                        <td>
```

```
                        Date of production
                    </td>
                </tr>
                    <s:iterator value="products" status="stat">
                        <tr>
                            <td>
                                <s:property value="name" />
                            </td>
                            <td>
                                $
                                <s:property value="price" />
                            </td>
                            <td>
                                <s:property value="dateOfProduction" />
                            </td>
                        </tr>
                </s:iterator>
            </table>
        </body>
</html>
```

发布运行应用程序，在浏览器中键入 http://localhost:8080/Struts2_Chapter04/AddProducts.jsp，出现如图 4-4（a）所示页面。

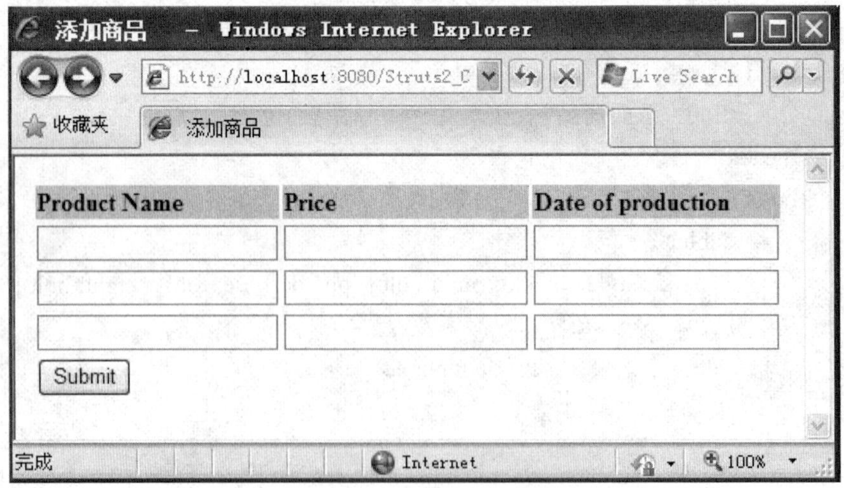

（a）

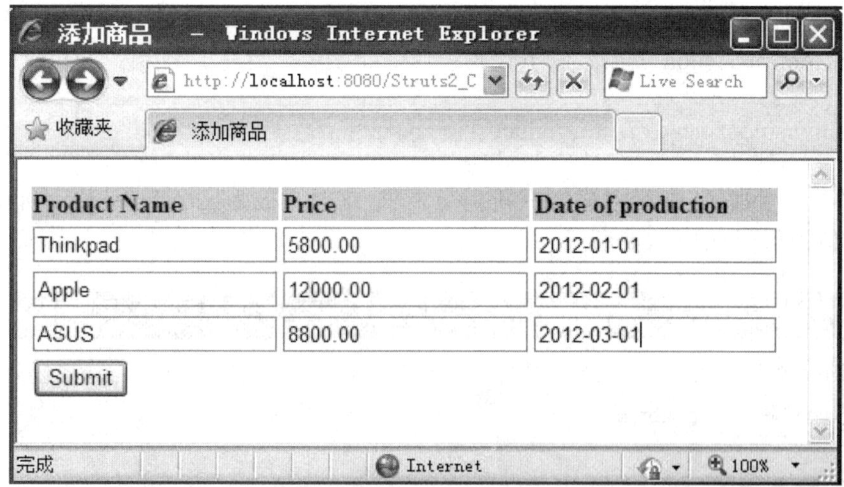

（b）添加信息页面

图 4-4　添加页面

按图 4-4（b）所示，填写表单，按“Submit”提交，出现图 4-5 所示页面。

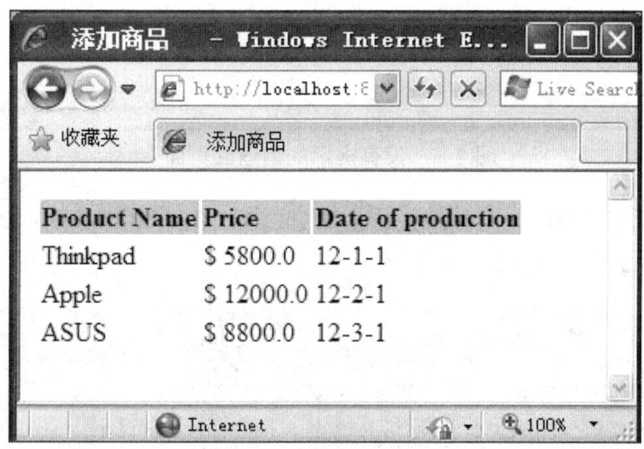

图 4-5　查看产品页面

4.3　转换错误处理

大家在运行以上的例子时，可能会出现填错日期或数字的情况，例如，在第一行的 Price 和 Date of production 中输入英文字母，然后按“Submit”提交。你会看到页面为空白，再看一下服务器的控制台输出，有如下语句：

```
警告：No result defined for action com.xtgj.struts2.chapter04.action.ProductConfirm and result input
```

它提示我们没有为 action 定义输入结果，所以，我们应该在源代码文件夹下的 struts.xml 中的 ProductConfirm Action 中加入以下代码：

```
<result name="input">/AddProducts.jsp</result>
```

重新加载应用程序，刷新浏览器重新提交请求，这时页面返回 AddProducts.jsp，格式错误的输入框的值被保留，如图 4-6 所示。

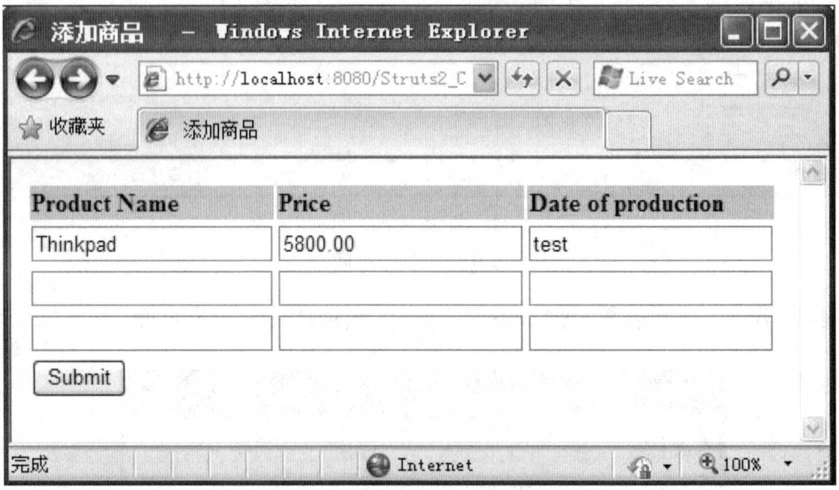

图 4-6 没有提示的错返回页面

当然，我们还可以在页面上加上错误提示信息，通过在 AddProducts.jsp 的"<body>"后，加入如示例代码 4-10 所示的代码可以实现。

示例代码 4-10 AddProducts.jsp 中加入错误提示信息

```
<%@ page contentType="text/html; charset=UTF-8"%>
<%@taglib prefix="s" uri="/struts-tags"%>
<html>
    <head>
        <title> 添加商品 </title>
    </head>
    <body>
        <div style="color: red">
            <s:fielderror />
        </div>
        // 省略部分代码
    </body>
</html>
```

刷新浏览器，重新提交请求，出现如图 4-7 所示页面。

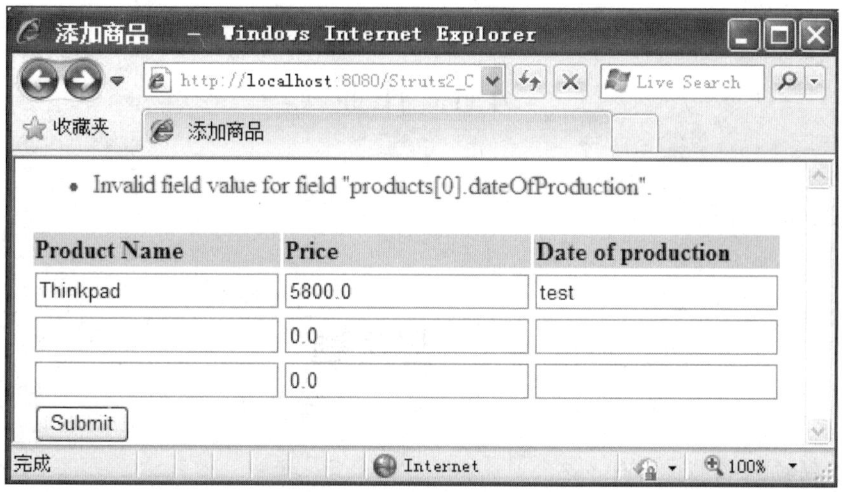

图 4-7　带提示的错返回页面

以上的功能都是通过 Struts 2 里的一个名为 conversionError 的拦截器（interceptor）实现的，它被注册到默认拦截器栈（default interceptor stack）中。Struts 2 在转换出错后，会将错误放到 ActionContext 中，conversionError 的作用是将这些错误封装为对应的项错误（field error），因此我们可以通过 <s:fielderror /> 来将其在页面上显示出来。另外，大家看第二和第三行的 Price 都被赋为 0.0 的值，而第一行则保留其错误值。这同样是 conversionError 的功劳——没有出错的行调用 products[index].price（默认值为 0.0），而出错的行则会被赋为页面所提交的错误值，这样可以提供更好的用户体验。

4.4　小结

- ✓ Struts 2 的转换器简化的 Web 应用程序的模型，为我们的编程带来极大的方便。
- ✓ 转换器分为全局类型转换器和特定类型转换器两种。
- ✓ 转换器的定义语法。

4.5　英语角

converter	转换器
locale	语言环境
OGNL	对象图导航语言

4.6 作业

回顾转换器的使用。

4.7 思考题

Struts 2 转换器的原理是什么？

4.8 学员回顾内容

1. 转换器的主要特性。
2. 转换器的组成结构。

第 5 章　Struts 2 表单数据校验

学习目标

✧ 了解数据校验的主要原理。
✧ 理解数据校验的组成部分。
✧ 掌握数据校验的使用。

课前准备

✧ Struts 2 数据校验的组成。
✧ Struts 2 数据校验的分类。

本章简介

本章主要介绍 Struts 2 表单数据校验主要原理，学习表单数据校验的实现方法，了解表单数据校验的校验框架以及现有的校验器的使用方法。

5.1　简述

表单是应用程序最简单的入口，对其传进来的数据，我们必须进行校验。转换是校验的基础，只有在数据被正确地转换成其对应的类型后，才可以对其取值范围进行校验。

Struts 2 中，我们可以实现对 action 的所有方法进行校验或者对 action 的指定方法进行校验。对于输入校验 Struts 2 提供了两种实现方法：

（1）采用手工编写代码实现。

（2）基于 XML 配置方式实现。

5.2　采用手工编写代码实现

请看下面的例子，逐步体会采用手工编写代码实现校验流程与其重要性。

5.2.1 对 action 中所有方法输入校验

本案例使用手工编写代码方式验证用户注册信息是否合法。

首先，从 action 开始。在 Struts 2 中通过 Action 重写 validate() 方法的实现，validate() 方法会校验 action 中所有与 execute() 方法签名相同的方法。当某个数据校验失败时，我们应该调用 addFieldError() 方法向系统的 fieldErrors 集合中添加校验失败信息（为了使用 addFieldError() 方法，action 可以继承 ActionSupport ），如果系统的 fieldErrors 包含失败信息，Struts 2 会将请求转发到名为 input 的 result。如示例代码 5-1 所示。

示例代码 5-1 UserAction.java 代码

```
package com.xtgj.struts2.chapter05.user;
import java.util.regex.Pattern;
import com.opensymphony.xwork2.ActionContext;
import com.opensymphony.xwork2.ActionSupport;
public class UserAction extends ActionSupport {
    private String username;
    private String mobile;
    private static final long serialVersionUID = -6328095689264546407L;
    public void validate() {
        if (this.mobile == null || "".equals(this.mobile.trim())) {
            this.addFieldError("mobile", " 手机号不能为空 ");
        } else {
            if (!Pattern.compile("^1[358]\\d{9}").matcher(this.mobile.trim())
                    .matches()) {
                this.addFieldError("mobile", " 手机号的格式不正确 ");
            }
        }
        if (this.username == null || this.username.equals("")) {
            this.addFieldError("username", " 用户名不能为空 ");
        }
    }
    public String update() {
        return "success";
    }
    public String delete() {
        return "success";
    }
    public String login() {
        return "success";
```

```
}
        public String regist() {
               return "success";
        }      public String getUsername() {
               return username;
        }
        public void setUsername(String username) {
               this.username = username;
        }
        public String getMobile() {
               return mobile;
        }
        public void setMobile(String mobile) {
               this.mobile = mobile;
        }
}
```

此 action 被划分在 user 模块中，user.xml 的具体配置如示例代码 5-2 所示。

示例代码 5-2　user.xml 代码

```
<?xml version="1.0" encoding="UTF-8"?>
<!DOCTYPE struts PUBLIC
       "-//Apache Software Foundation//DTD Struts Configuration 2.3//EN"
       "http://struts.apache.org/dtds/struts-2.3.dtd">
<struts>
       <package name="user" namespace="/user" extends="struts-default">

              <global-results>
                     <result name="msg">msg.jsp</result>
              </global-results>
              <action   name="user_*"   class="com.xtgj.struts2.chapter05.user.UserAction"
method="{1}">
                     <result name="success">success.jsp</result>
                     <result name="input">register.jsp</result>
              </action>
       </package>
</struts>
```

struts.xml 中的配置如示例代码 5-3 所示。

示例代码 5-3 struts.xml

```
<?xml version="1.0" encoding="UTF-8" ?>
<!-- 指定 Struts 2 配置文件的 DTD 信息 -->
<!DOCTYPE struts PUBLIC
    "-//Apache Software Foundation//DTD Struts Configuration 2.3//EN"
    "http://struts.apache.org/dtds/struts-2.3.dtd">
<!-- struts 是 Struts 2 配置文件的根元素 -->
<struts>
    <constant name="struts.custom.i18n.resources"
        value="globalMessages" />
    <include file="user.xml" />
</struts>
```

用户注册页面 register.jsp 代码如示例代码 5-4 所示。

示例代码 5-4 register.jsp 代码

```
<%@ page language="java" import="java.util.*" pageEncoding="UTF-8"%>
<%@taglib prefix="s" uri="/struts-tags"%>
<!DOCTYPE HTML PUBLIC "-//W3C//DTD HTML 4.01 Transitional//EN">
<html>
    <head>
        <title>My JSP 'register.jsp' starting page</title>
    </head>

    <body>
        <s:fielderror />
        <table width="100%" border="0" cellspacing="0" cellpadding="4">
            <tr>
                <td bgcolor="#000099">
                    <table width="100%" border="0" cellspacing="0" cellpad-
ding="4">
                        <tr>
                            <td width="100%">
                                <font color="#CCCCCC">  <font
color="#FFFFFF"> 用户注册 </font>
```

```
                </font>
              </td>
            </tr>
          </table>
        </td>
      </tr>
      <tr>
        <td width="100%" bgcolor="#EAEAEA" colspan="2">
          <form name="Name" action="user_regist.action" method="post">
            <p>
              <label for="textfield">
                用户名
              </label>
                <input type="text" name="username" id="text-
field">
            </p>
            <p>
              <label for="textfield2">
                手     机
              </label>
              <input type="text" name="mobile" id="textfield2">
            </p>
            <p>
              <input type="submit" name="Submit" value=" 提交 ">
            </p>
            <p>
               
            </p>
            </form>
          </td>
        </tr>
      </table>
    </body>
  </html>
```

页面中“<s:fielderror/>”是错误提示代码，若注册信息验证失败，页面仍停留在注册页。在该页中可以通过 <s:fielderror/> 显示失败信息。发布运行应用程序，在浏览器中键入“http://localhost: 8080/Struts 2_Chapter05 /user/register.jsp”，显示注册页面如图 5-1 所示。

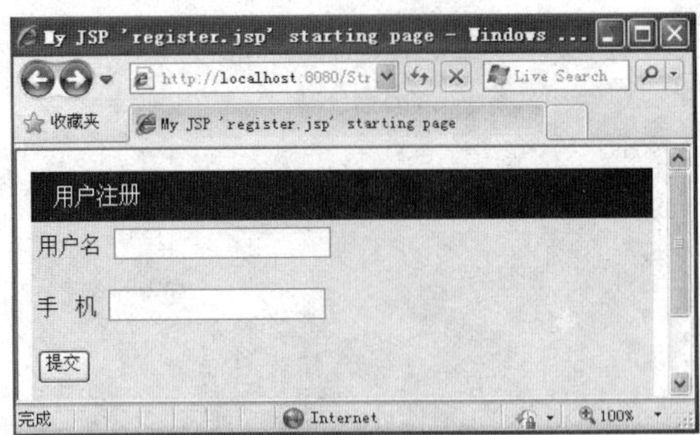

图 5-1 控制台输出结果

若未输入信息就点击了提交按钮，则显示如图 5-2 所示的结果。

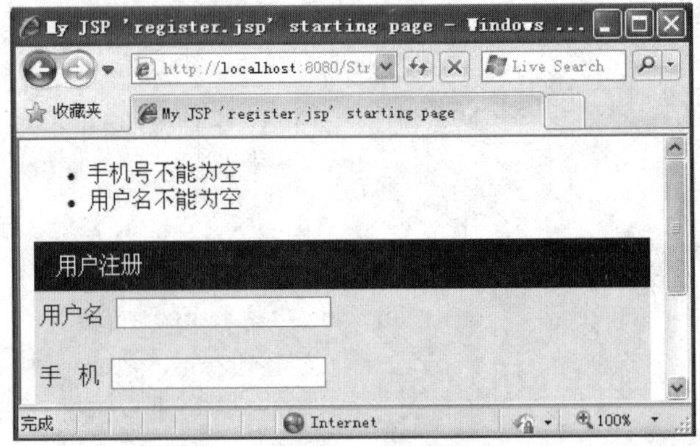

图 5-2 手机号和用户名不能为空

输入错误格式的手机号，将显示如图 5-3 所示的效果。

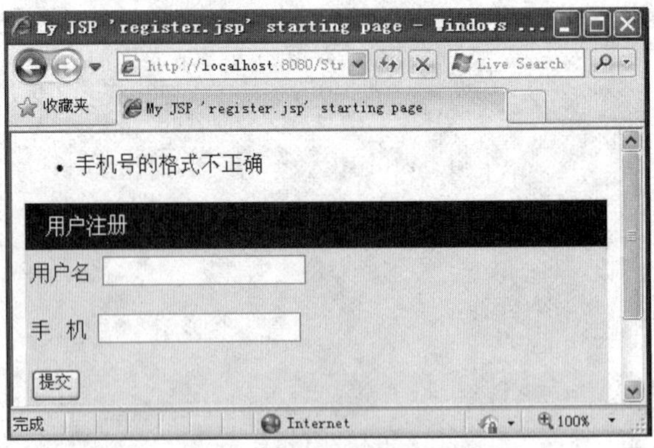

图 5-3 手机号格式不正确

5.2.2 对 action 指定方法输入校验

在上文的介绍中通过重写 validate() 方法，实现校验 action 中所有与 execute() 方法名相同的方法。但是，并不是所有的方法都需要校验，如果只想对特定的方法进行校验，可以通过

validateXxx() 方法实现，validateXxx() 只会校验 action 中方法名为 Xxx 的方法，这是利用 Java 的反射机制实现的。其中 Xxx 的第一个字母要大写。当某个数据校验失败时，应调用 addFieldError() 方法往系统的 fieldErrors 添加校验失败信息（为了使用 addFieldError() 方法，action 可以继承 ActionSupport），如果系统的 fieldErrors 包含失败信息，Struts 2 会将请求转发到名为 input 的 result。

例如，上述案例中，如果仅对注册业务或用户信息修改业务实行校验功能，而不对其他的删除、查询等业务进行校验就可实现，如示例代码 5-5 所示。

示例代码 5-5　修改后的 UserAction.java 代码

```
package com.xtgj.struts2.chapter05.user;
import java.util.regex.Pattern;
import com.opensymphony.xwork2.ActionContext;
import com.opensymphony.xwork2.ActionSupport;
public class UserAction extends ActionSupport {
        private String username;
        private String mobile;
        private static final long serialVersionUID = -6328095689264546407L;
        public void checkmethod() {
                if (this.mobile == null || "".equals(this.mobile.trim())) {
                        this.addFieldError("mobile", " 手机号不能为空 ");
                } else {
                        if (!Pattern.compile("^1[358]\\d{9}").matcher(this.mobile.trim())
                                        .matches()) {
                                this.addFieldError("mobile", " 手机号的格式不正确 ");
                        }
                }
                if (this.username == null || this.username.equals("")) {
                        this.addFieldError("username", " 用户名不能为空 ");
                }
        }
        public void validateRegist() {
                checkmethod();
        }
        public void validateUpdate() {
                checkmethod();
        }
        public String update() {
                return "success";
```

```
    }
    public String delete() {
        return "success";
    }
    public String login() {
        return "success";
    }
    public String regist() {
        return "success";
    }
    public String getUsername() {
        return username;
    }
    public void setUsername(String username) {
        this.username = username;
    }
    public String getMobile() {
        return mobile;
    }
    public void setMobile(String mobile) {
        this.mobile = mobile;
    }
}
```

仅修改 UserAction 类，其他的结构不改变，这样的情况下，只有“user_regist”和“user_update”请求会实现校验，并对特定的“regist”方法和“update”方法校验。

5.3 数据校验工作方式

数据校验需要经过下面几个步骤：

（1）类型转换器对请求参数执行类型转换，并把转换后的值赋给 action 中的属性。

（2）如果在执行类型转换的过程中出现异常，系统会将异常信息保存到 ActionContext，conversionError 拦截器将异常信息添加到 fieldErrors 里。不管类型转换是否出现异常，都会进入第 3 步。

（3）系统通过反射技术先调用 action 中的 validateXxx() 方法，Xxx 为方法名。

（4）再调用 action 中的 validate() 方法。

（5）经过上述 4 步，如果系统中的 fieldErrors 存在错误信息（即存放错误信息的集合的

size 大于 0),系统自动将请求转发至名称为 input 的视图。如果系统中的 fieldErrors 没有任何错误信息,系统将执行 action 中的处理方法。处理流程如图 5-4 所示。

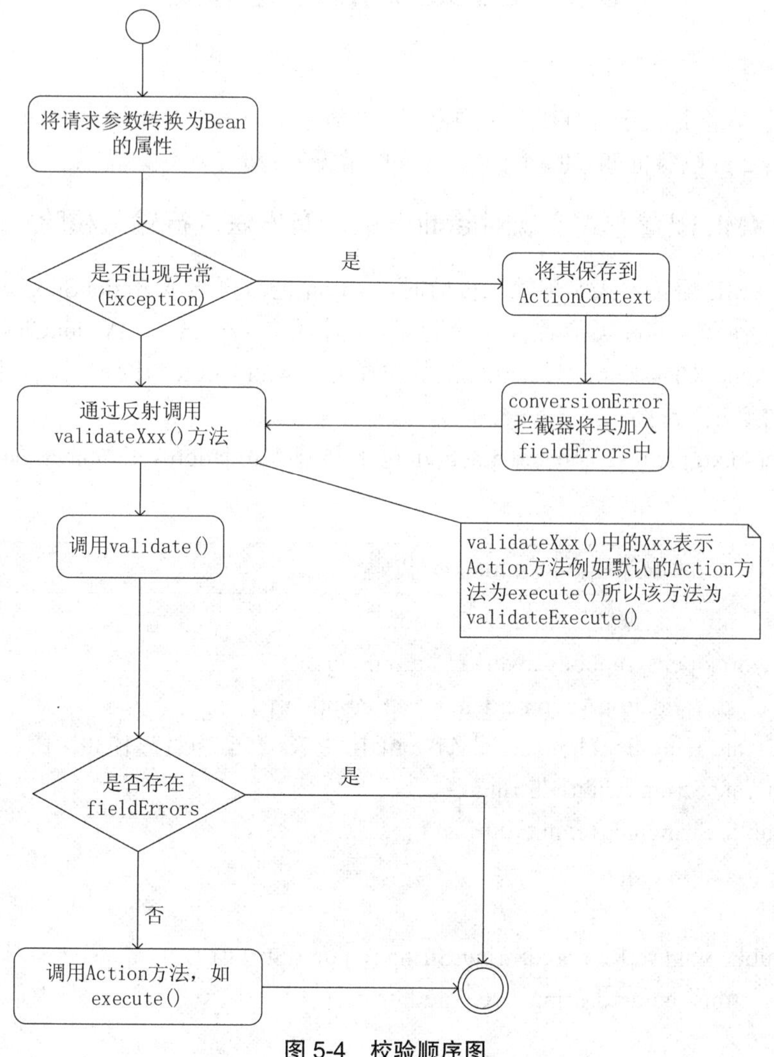

图 5-4　校验顺序图

看到这里可能大家会提出疑问:“这么多地方可以校验表单数据,到底应该在哪里做呢?”有选择是好事,但选择的过程往往是痛苦的,且让人不知所措。如果大家参照以下几点建议,相信会比较容易地做出正确的选择。

对于需要转换的数据,通常做法是在转换的时候做格式的校验,在 action 中的校验方法中校验取值。假如用户填错了格式,我们可以通过在资源文件配置 invalid.fieldvalue.xxx(xxx 为属性名)来提示用户正确的格式,不同的出错阶段显示不同的信息,具体做法请参考上面的例子。

至于用 validate() 还是 validateXxx(),推荐使用 validate()。原因是 validateXxx() 使用了反射,相对来说性能稍差,而 validate() 则是通过接口 com.opensymphony.xwork2.Validateable 调用。当然如果你的表单数据取值是取决于特定 action 方法,则应该使用 validateXxx()。

5.4 Struts 2 的校验框架

上一节的内容都是关于如何编程实现校验，这部分工作人都是单调重复的。更多情况下，我们使用 Struts 2 的校验框架，通过配置实现一些常见的校验。

5.4.1 基于 XML 配置方式实现对 action 的所有方法进行输入校验

使用基于 XML 配置方式实现输入校验时，action 也需要继承 ActionSupport，并且提供校验文件，校验文件和 action 类放在同一个包下，文件的取名格式为：ActionClassName-validation.xml，其中 ActionClassName 为 action 的简单类名，-validation 为固定写法。以下案例主要实现文本信息“必填”的验证。

首先，在 com.xtgj.struts2.chpater05.action 包下新建 ValidationAction.java，如示例代码 5-6 所示。

示例代码 5-6 ValidationAction.java 代码

```
package com.xtgj.struts2.chpater05.action;
import com.opensymphony.xwork2.ActionSupport;
public class ValidationAction extends ActionSupport {
    private static final long serialVersionUID = 7862562341133193022L;
    private String reqiuredString;
    public String getReqiuredString() {
        return reqiuredString;
    }
    public void setReqiuredString(String reqiuredString) {
        this.reqiuredString = reqiuredString;
    }
    @Override
    public String execute() {
        return SUCCESS;
    }
}
```

然后，配置 action，如示例代码 5-7 所示。

示例代码 5-7　struts.xml 代码片段

```
<action name ="ValidationAction" class ="com.xtgj.struts2.chpater05.action.Validation-
Action">
        <result>/Output.jsp</result>
        <result name ="input">/Input.jsp</result >
</action>
```

接着，创建信息录入页面 Input.jsp，如示例代码 5-8 所示。

示例代码 5-8　Input.jsp 代码

```
<%@ page  contentType = " text/html; charset=UTF-8"%>
<%@ taglib prefix = "s" uri ="/struts-tags" %>
<html>
<head>
    <!-- 此标志的作用是引入 Struts 2.0 的常用的 Javascript 和 CSS -->
    <s:head/>
</head>
<body>
    <s:form action ="ValidationAction" >
        <s:textfield name ="reqiuredString" label ="Required String" />
        <s:submit />
    </s:form >
</body>
</html>
```

编写校验成功后的信息输出页面 Output.jsp，如示例代码 5-9 所示。

示例代码 5-9　Output.jsp 代码

```
<%@page  contentType = " text/html; charset=UTF-8"%>
<%@taglib prefix = "s" uri = "/struts-tags"%>
<html>
<head>
    <title> output </title>
</head>
<body>
    Required String: <s:property value ="reqiuredString"/>
</body>
</html>
```

接下来，在 com.xtgj.struts2.chpater05.action 包下创建 ValidationAction 的校验配置文件 xx-validation.xml（Xxx 为 Action 的类名），在本例中该文件名为：ValidationAction-validation.xml。如示例代码 5-10 所示。

示例代码 5-10　ValidationAction-validation.xml 代码

```
<?xml version="1.0" encoding="UTF-8"?>
<!DOCTYPE validators PUBLIC "-//OpenSymphony Group//XWork Validator 1.0//
EN" "http://www.opensymphony.com/xwork/xwork-validator-1.0.dtd" >
<validators>
    <field name="reqiuredString">
        <field-validator type="requiredstring">
            <param name="trim">true</param>
            <message>This string is required</message>
        </field-validator>
    </field>
</validators>
```

<field> 指定 action 中要校验的属性，<field-validator> 指定校验器，上面指定的校验器 requiredstring 是由系统提供的，系统提供了能满足大部分验证需求的校验器，这些校验器的定义可以在 xwork-2.x.jar 中的 com.opensymphony.xwork2.validator.validators 下的 default.xml 中找到。

<message> 为校验失败后的提示信息，如果需要国际化，可为 message <message> 指定 key 属性，key 的值为资源文件中的 key。

在这个校验文件中，对 action 中字符串类型的 reqiuredString 属性进行验证，首先要求调用 trim() 方法去掉空格，然后判断用户名是否为空。

发布应用程序，在地址栏中键入“http://localhost:8080/Struts2_Chapter05/Input.jsp”，出现如图 5-5 所示页面。

图 5-5　Input.jsp 界面

直接点击“Submit”提交表单，出现图 5-6 所示的页面。

图 5-6　错误提示

在 Required String 中填写“No empty”，转到 Output.jsp 页面，如图 5-7 所示。

图 5-7　Output.jsp 输出界面

通过上面的例子，大家可以看到使用该校验框架十分简单方便。不过，上例还有两点不足：

（1）还没有国际化错误消息。

（2）没有实现客户端的校验。

当然，要完善以上不足，对于 Struts 2 来说，只是小菜一碟。在 Xxx-validation.xml 文件中的 <message> 元素中加入 key 属性；在 Input.jsp 中的 <s:form> 标志中加入 validate="true" 属性，就可以再用 JavaScript 在客户端校验数据。

5.4.2　基于 XML 配置方式对指定 action 方法实现输入校验

当校验文件的取名为 ActionClassName-validation.xml 时，会对 action 中的所有处理方法实施输入验证。如果只需要对 action 中的某个 action 方法实施校验，那么，校验文件的取名应为：ActionClassName-ActionName-validation.xml，其中， ActionClassName 为 action 类的名称，ActionName 为 struts.xml 文件中 action 的名称。

以上述示例为例，可以在 com.xtgj.struts2.chpater05.action 包下创建 ValidationAction 的校验配置文件 ValidationAction-ValidationAction_check-validation.xml，如示例代码 5-11 所示。

示例代码 5-11　ValidationAction-ValidationAction_check-validation.xml 代码

```
<?xml version="1.0" encoding="UTF-8"?>
<!DOCTYPE validators PUBLIC "-//OpenSymphony Group//XWork Validator 1.0//
EN" "http://www.opensymphony.com/xwork/xwork-validator-1.0.dtd" >
<validators>
    <field name="reqiuredString">
        <field-validator type="requiredstring">
            <param name="trim">true</param>
            <message>This string is required</message>
        </field-validator>
    </field>
</validators>
```

修改 struts.xml 配置信息，如示例代码 5-12 所示。

示例代码 5-12　struts.xml 代码

```
<?xml version="1.0" encoding="UTF-8" ?>
<!DOCTYPE struts PUBLIC
    "-//Apache Software Foundation//DTD Struts Configuration 2.0//EN"
    "http://struts.apache.org/dtds/struts-2.0.dtd">
<struts>
    <constant name="struts.custom.i18n.resources"
        value="globalMessages" />
    <include file="user.xml" />
    <package name="valid" extends="struts-default">
        <action name="ValidationAction_*"
            class="com.xtgj.struts2.chpater05.action.ValidationAction" meth-
od="{1}">
            <result>/Output.jsp</result>
            <result name="input">/Input.jsp</result>
        </action>
    </package>
</struts>
```

修改 ValidationAction.java，在该类中添加方法 check()，如示例代码 5-13 所示。

示例代码 5-13　check() 方法代码

```
package com.xtgj.struts2.chpater05.action;
import com.opensymphony.xwork2.ActionSupport;
public class ValidationAction extends ActionSupport {
    // 省略部分代码
    public String execute() {
        return SUCCESS;
    }
    public String check() {
        System.out.println("Action 中的特定方法 ");
        return SUCCESS;
    }
}
```

修改 Input.jsp 页面的表单 action 请求为“ValidationAction_check”，这时验证框架仅对 check() 方法有效。实现效果同上。

5.4.3　配置文件查找顺序

在上述例子中，通过创建 ValidationAction-validation.xml 来配置表单校验。Struts 2 的校验框架自动会读取该文件，但这样就会引出一个问题——如果我的 action 继承其他的 Action 类，而这两个 action 类都需要对表单数据进行校验，是否会在子类的配置文件（Xxx-validation.xml）中复制父类的配置吗？

答案是否定的，因为 Struts 2 的校验框架有特定的配置文件查找顺序，校验框架按照自上而下的顺序在类层次查找配置文件。假设以下条件成立：

（1）接口 Animal。

（2）接口 Quadraped 扩展了 Animal。

（3）类 AnimalImpl 实现了 Animal。

（4）类 QuadrapedImpl 扩展了 AnimalImpl 实现了 Quadraped。

（5）类 Dog 扩展了 QuadrapedImpl。

如果 Dog 要被校验，框架方法会查找下面的配置文件（其中别名是 Action 在 struts.xml 中定义的别名）：

（1）Animal-validation.xml。

（2）Animal- 别名 -validation.xml。

（3）AnimalImpl-validation.xml。

（4）AnimalImpl- 别名 -validation.xml。

（5）Quadraped-validation.xml。

（6）Quadraped- 别名 -validation.xml。

（7）QuadrapedImpl-validation.xml。

（8）QuadrapedImpl- 别名 -validation.xml。

（9）Dog-validation.xml。

（10）Dog- 别名 -validation.xml。

5.4.4 已有的校验器

Struts 2 已经实现很多常用的校验了，示例代码 5-14 所示是在 xwork-core-2.1.6.jar 的 default.xml 中注册的校验器。

示例代码 5-14 xwork-core-2.1.6.jar 的 default.xml 代码

```
<?xml version="1.0" encoding="UTF-8"?>
<!DOCTYPE validators PUBLIC
        "-//OpenSymphony Group//XWork Validator Config 1.0//EN"
        "http://www.opensymphony.com/xwork/xwork-validator-config-1.0.dtd">
<!-- START SNIPPET: validators-default -->
<validators>
    <validator name="required" class="com.opensymphony.xwork2.validator.validators.RequiredFieldValidator"/>
    <validator name="requiredstring" class="com.opensymphony.xwork2.validator.validators.RequiredStringValidator"/>
    <validator name="int" class="com.opensymphony.xwork2.validator.validators.IntRangeFieldValidator"/>
    <validator name="long" class="com.opensymphony.xwork2.validator.validators.LongRangeFieldValidator"/>
    <validator name="short" class="com.opensymphony.xwork2.validator.validators.ShortRangeFieldValidator"/>
    <validator name="double" class="com.opensymphony.xwork2.validator.validators.DoubleRangeFieldValidator"/>
    <validator name="date" class="com.opensymphony.xwork2.validator.validators.DateRangeFieldValidator"/>
    <validator name="expression" class="com.opensymphony.xwork2.validator.validators.ExpressionValidator"/>
    <validator name="fieldexpression" class="com.opensymphony.xwork2.validator.validators.FieldExpressionValidator"/>
    <validator name="email" class="com.opensymphony.xwork2.validator.validators.EmailValidator"/>
    <validator name="url" class="com.opensymphony.xwork2.validator.validators.URLValidator"/>
```

```
        <validator name="visitor" class="com.opensymphony.xwork2.validator.validators.VisitorFieldValidator"/>
        <validator name="conversion" class="com.opensymphony.xwork2.validator.validators.ConversionErrorFieldValidator"/>
        <validator name="stringlength" class="com.opensymphony.xwork2.validator.validators.StringLengthFieldValidator"/>
        <validator name="regex" class="com.opensymphony.xwork2.validator.validators.RegexFieldValidator"/>
        <validator name="conditionalvisitor" class="com.opensymphony.xwork2.validator.validators.ConditionalVisitorFieldValidator"/>
    </validators>
    <!-- END SNIPPET: validators-default -->
```

5.5　小结

✓ 使用校验框架既可以方便地实现表单数据校验，又能够将校验与 action 分离。

✓ 在使用校验框架时，请不要忘记在资源文件加入 invalid.fieldvalue.Xxx 字符串，显示适合的类型转换出错信息，或者使用 Conversion 校验器。

5.6　英语角

register	寄存器
conversion	转换
invalid	无效
quadraped	四足

5.7　作业

1. 回顾 Struts 2 表单数据校验的过程。
2. 将本章中的第一个案例改为验证框架实现方式。

5.8 思考题

Struts 2 表单数据校验的种类有什么？

5.9 学员回顾内容

1. 数据校验的主要原理。
2. 理解数据校验的组成部分。

第 6 章　Struts 2 拦截器

学习目标

✧ 了解 Struts 2 拦截器的意义。
✧ 理解 Struts 2 拦截器的工作原理。
✧ 掌握 Struts 2 拦截器的应用。

课前准备

了解拦截器的定义。

本章简介

本章主要介绍拦截器的工作原理，学习如何配置拦截器以及拦截器的意义，了解如何使用自定义拦截器。

6.1　理解拦截器

Interceptor（以下译为拦截器）是 Struts 2 的一个强有力的工具，有许多功能（feature）都是构建于它之上，如国际化、转换器、校验等。

在 Action() 执行前后 Interceptor 都要被执行。框架的大部分核心功能（包括类型转化，防止双重提交等）都是借助拦截器实现的。所有的拦截器都是插件式的，你可以为你的 action 精确地设置需要的拦截器。

6.1.1　拦截器的工作原理

大部分时候，拦截器方法都是通过代理的方式来调用的。Struts 2 的拦截器实现相对简单。当请求到达 Struts 2 的 StrutsPrepareAndExecuteFilter 时，Struts 2 会查找配置文件，并根据其配置实例化相对的拦截器对象，然后串成一个列表（list），最后一个一个地调用列表中的拦截器，如图 6-1 所示。

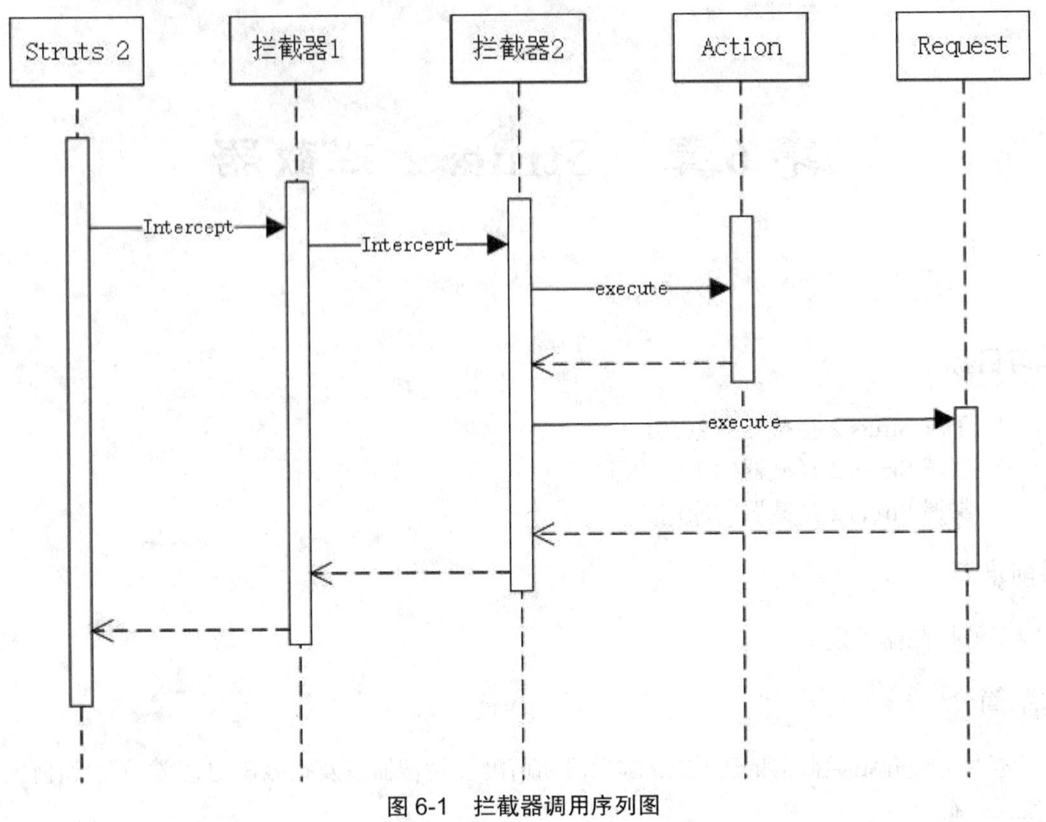

图 6-1 拦截器调用序列图

6.1.2 拦截器的意义

拦截器，在 AOP（Aspect-Oriented Programming）中用于在某个方法或字段被访问之前，进行拦截，然后在之前或之后加入某些操作。拦截是 AOP 的一种实现策略。

在 WebWork 中文文档中的解释为——拦截器是动态拦截 action 调用的对象。它提供了一种机制可以使开发者能够定义在一个 action 执行前阻止其执行的程序。同时也提供了一种可以提取 action 中可重用的部分的方式。

谈到拦截器，还有一个词大家应该知道——拦截器链（Interceptor Chain），在 Struts 2 中称为拦截器栈（Interceptor Stack）。拦截器链就是将拦截器按一定的顺序联结成一条链。在访问被拦截的方法或字段时，拦截器链中的拦截器就会按其之前定义的顺序被调用。

6.2 配置拦截器

Struts 2 已经为您提供丰富多样、功能齐全的拦截器。大家可以到 struts-2.3.15.3-all.zip 或 struts2-core-2.3.15.3jar 包的 struts-default.xml 文件中查看关于默认的拦截器与拦截器栈的配置。

6.2.1　定义拦截器

在 struts.xml 文件中定义拦截器，如示例代码 6-1 所示。

示例代码 6-1　拦截器的配置代码

```
<package name="my" extends="struts-default" namespace="/manage">
        <interceptors>
        <!-- 定义拦截器 -->
        <interceptor name=" 拦截器名 " class=" 拦截器实现类 "/>
        <!-- 定义拦截器栈 -->
        <interceptor-stack name=" 拦截器栈名 ">
            <interceptor-ref name=" 拦截器一 "/>
            <interceptor-ref name=" 拦截器二 "/>
        </interceptor-stack>
        </interceptors>
        ......
</package>
```

在 struts.xml 配置文件中，如果要定义拦截器，需要为拦截器类指定一个拦截器名。拦截器的定义使用 <interceptor/> 元素来实现，该元素定义格式一般如示例代码 6-2。

示例代码 6-2　定义拦截器元素

```
<!-- 定义拦截器或者拦截器栈 -->
<interceptor name=" 拦截器 A" class=" 拦截器实现类 A"/>
```

这是一个比较简单的拦截器定义格式，开发者在进行项目开发时，一般只会用到以上拦截器定义。<interceptor/> 元素还可以定义 <param/> 参数子元素，用来对拦截器的参数初始化，如示例代码 6-3 所示。

示例代码 6-3　为拦截器初始化参数

```
<!-- 定义拦截器或者拦截器栈 -->
<interceptor name=" 拦截器 A" class=" 拦截器实现类 A"/>
    <!-- 定义拦截器参数 -->
    <param name=" 参数 A"> 参数 A 初始化值 </param>
    <param name=" 参数 B"> 参数 B 初始化值 </param>
    … …
</interceptor>
… …
```

在 Struts 2 配置文件中，除了定义拦截器，还可以定义拦截器栈。拦截器栈就是由多个拦截器组成的一个拦截器组，来实现一个特定拦截功能。其使用方法和功能与拦截器类似。

例如，可以将用户安全认证拦截器、日志记录拦截器、密码加密拦截器和其他相关拦截器组成一个拦截器栈，在一个业务控制器 action 执行之前使用，比如用户注册 action，在执行该 action 前来执行该拦截器栈，进行一系列公共功能方面的处理。

在配置文件中定义拦截器栈，使用 <interceptor-stack/> 元素，该元素中需要定义子元素 <interceptor-ref/>，即该拦截器栈所包含的拦截器引用，如示例代码 6-4 所示。

示例代码 6-4　拦截器栈的定义

```
<!-- 定义拦截器 -->
<interceptors>
    … …
    <interceptor name="exception" class="com.opensymphony.xwork2.interceptor.ExceptionMappingInterceptor"/>
    <interceptor name="prepare" class="com.opensymphony.xwork2.interceptor.PrepareInterceptor"/>
    <interceptor name="servletConfig" class="com.opensymphony.xwork2.interceptor.ServletConfigInterceptor"/>
    <interceptor name="checkbox" class="com.opensymphony.xwork2.interceptor.CheckboxInterceptor"/>
    … …
<!-- 定义一个名为 "basicStack" 的拦截器栈，其中包含了多个拦截器引用 -->
    <interceptor-stack name="basicStack">
    <interceptor-ref name="exception" />
    <interceptor-ref name="servletConfig" />
    <interceptor-ref name="prepare" />
    <interceptor-ref name="checkbox" />
    <interceptor-ref name="params" />
    <interceptor-ref name="conversionError"/>
</interceptor-stack>
… …
</interceptors>
```

在以上代码所示的拦截器栈定义中，包含了多个拦截器引用。引用的拦截器在拦截器栈外已经定义过。例如，拦截器“exception”已经在前面 <interceptor/> 中定义过。

在拦截器栈定义的 <interceptor-ref/> 元素中，name 除了可以是一个拦截器名称，同时也可以是一个其他拦截器栈的名称，即该拦截器栈包含了其他拦截器栈，同时包含了该拦截器栈中所有的拦截器定义。例如，可以定义一个拦截器栈，如示例代码 6-5 所示。

示例代码 6-5　拦截器栈的定义

```
<!-- 定义一个名为 "MyStack" 的拦截器栈,其中包含了一个拦截器栈 "basicStack" -->
<interceptor-stack name="MyStack" />
    <interceptor-ref name="basicStack" />
    <interceptor-ref name="MyInterceptor" />
    ... ...
</interceptor-stack>
... ...
```

从以上代码可以看到,在拦截器栈 MyStack 的定义中,包含了一个拦截器 MyInterceptor 和一个拦截器栈 basicStack 的引用。这意味着,拦截器栈 MyStack 包含了拦截器栈 basicStack 所有的拦截器引用,例如拦截器 exception。

拦截器栈在功能上与拦截器没有任何区别,都是完成一个特定的拦截任务。Struts 2 框架可以在定义拦截器和使用拦截器时来初始化拦截器参数。一般在定义拦截器时使用 <param/> 初始化参数值是拦截器的默认参数;而在使用拦截器时(一般在 action 定义中使用拦截器)会动态指定拦截器参数值。拦截器参数初始化,如示例代码 6-6 所示。

示例代码 6-6　初始化拦截器参数

```
<!-- 定义一个名为 "MyStack" 的拦截器栈,其中包含了一个拦截器栈 "basicStack" -->
<interceptor-stack name="MyStack"/>
    <interceptor-ref name="basicStack"/>
    <interceptor-ref name="MybasicStack" />
    <!-- 初始化拦截器参数 -->
    <param name=" 参数 A"> 参数 A 初始化值 </param>
    ... ...
</interceptor-stack>
... ...
<action name="Reg" class="chapter05.Reg">
    <result name="success">/chapter05/success.jsp</result>
    <result name="input">/chapter05/reg.jsp</result>
    <!-- 引用默认拦截器 -->
    <interceptor-ref name="defaultStack"/>
    <interceptor-ref name="MyInterceptor"/>
    <!-- 在使用拦截器时指定拦截器参数 -->
    <param name=" 参数 A"> 参数 A 初始化值 </param>
    ... ...
</action>
```

注意:如果在定义拦截器时和使用拦截器时指定了不同的参数值,那么使用拦截器时指定的参数值会覆盖定义拦截器时指定的参数值;定义拦截器时指定的参数值为该拦截器默认参数值。

6.2.2 使用拦截器

如果您想要使用拦截器,只需要在应用程序 struts.xml 文件中通过"<include file="struts-default.xml" />"将 struts-default.xml 文件包含进来,并继承其中的 struts-default 包(package),最后在定义 action 时,使用"<interceptor-ref name="xx" />"引用拦截器或拦截器栈(Interceptor stack)。一旦继承了 struts-default 包(package),所有 action 都会调用拦截器栈 ——defaultStack。当然,在 action 配置中加入"<interceptor-ref name="xx" />"可以覆盖 defaultStack。

下面是关于拦截器 timer 使用的例子。

首先,新建 action 类 com.xtgj.struts2.chapter06.action.TimerInterceptorAction.java,如示例代码 6-7 所示。

示例代码 6-7 TimerInterceptorAction.java 代码

```
package com.xtgj.struts2.chapter06.action;

import com.opensymphony.xwork2.ActionSupport;

public class TimerInterceptorAction extends ActionSupport {
    private static final long serialVersionUID = 5309310554138672331L;
    @Override
    public String execute() {
        try {
            // 模拟耗时的操作
            Thread.sleep(500);
        } catch (Exception e) {
            e.printStackTrace();
        }
        return SUCCESS;
    }
}
```

配置 action,名为 Timer,配置文件如示例代码 6-8 所示。

示例代码 6-8　在 struts.xml 中配置 Timer Action

```
<?xml version="1.0" encoding="UTF-8"?>
<!DOCTYPE struts PUBLIC
    "-//Apache Software Foundation//DTD Struts Configuration 2.3//EN"
    "http://struts.apache.org/dtds/struts-2.3.dtd">
<struts>
    <include file="struts-default.xml" />
    <package name="InterceptorDemo" extends="struts-default">
        <action name="Timer"
            class="com.xtgj.struts2.chapter06.action.TimerInterceptorAction">
            <interceptor-ref name="timer" />
            <result>/success.jsp</result>
        </action>
    </package>
</struts>
```

新建 Timer.jsp 页面，如示例代码 6-9 所示。

示例代码 6-9　Timer.jsp 页面代码

```
<%@ page language="java" pageEncoding="gbk"%>
<!DOCTYPE HTML PUBLIC "-//W3C//DTD HTML 4.01 Transitional//EN">
<html>
    <head>
        <title>Timer</title>
    </head>
    <body>
        <center>
            <b> 用户登录 </b>
            <br>
            <form action="Timer.action" method="post">
                <input type="submit" value=" 提交 " />
            </form>
        </center>
    </body>
</html>
```

新建 success.jsp 页面，如示例代码 6-10 所示。

示例代码 6-10 success.jsp 页面代码

```
<%@ page language="java" pageEncoding="gbk"%>
<!DOCTYPE HTML PUBLIC "-//W3C//DTD HTML 4.01 Transitional//EN">
<html>
    <head>
        <title>success</title>
    </head>
    <body>
        <center>
            <h1>
                已完成操作！
            </h1>
        </center>
    </body>
</html>
```

发布运行程序，在 Timer.jsp 页面点击提交按钮成功跳转到 success.jsp 页面后，服务器的后台输出如图 6-2 所示。

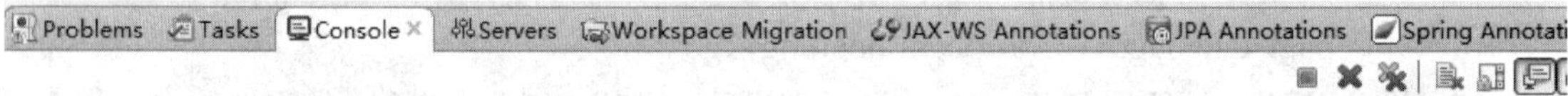


```
2016-02-10 14:21:09 com.opensymphony.xwork2.util.logging.commons.CommonsLogger info
信息: Excuted action[//Timer!execute] took 1531 ms.
```

图 6-2 服务器输出

在你的环境中执行 Timer!execute 的耗时，可能与上述的时间有些不同，这取决于你 PC 的性能。当你重新请求 Timer.action 时，如图 6-3 所示。

```
Problems  Tasks  Console  Servers  Workspace Migration  JAX-WS Annotations  JPA Annotations  Spring Annotatior
<terminated> dd [Java Application] D:\myeclipse\binary\com.sun.java.jdk7.win32.x86_64_1.7.0.u45\bin\javaw.exe (2017年2月10日 下午2:22:48)
2016-02-10 14:23:06 com.opensymphony.xwork2.util.logging.commons.CommonsLogger info
信息: Excuted action[//Timer!execute] took 500 ms.
```

图 6-3 服务器输出

你会发现两次执行相同的程序，但是所用的时间却不同，这是什么原因呢？其实原因是第一次加载 Timer 时，需要进行一定的初始工作。

这正是我们期待的结果。上述例子演示了拦截器 Timer 的用途——用于显示执行某个 action 方法的耗时，在我们做一个粗略的性能调试时，这相当有用。

6.2.3　默认拦截器

配置文件在定义一个包时，可以指定默认拦截器，例如 struts.xml 中配置的包 chapter06，如示例代码 6-11 所示。

示例代码 6-11　在 struts.xml 中配置 chapter06 包

```
<package name="chapter06" extends="struts-default">
        <interceptors>
            自定义拦截器
            <interceptor name="simpleInterceptor"
                class="com.xtgj.struts2.chapter06.intercepter.SimpleInterceptor" />
        </interceptors>
        <action name="Reg"
            class="com.xtgj.struts2.chapter06.action.Reg">
            <result name="success">/success.jsp</result>
            <result name="input">/reg.jsp</result>
            引用拦截器
            <interceptor-ref name="defaultStack" />
            <interceptor-ref name="simpleInterceptor" />
        </action>
</package>
```

该配置文件中定义了包 chapter06，chapter06 继承了包 struts-default, 同时也继承了 struts-default 包下的默认拦截器。示例代码 6-12 是系统包 struts-default 的定义。

示例代码 6-12　Struts 2 系统包 struts-default 的定义

```
<!DOCTYPE struts PUBLIC
    "-//Apache Software Foundation//DTD Struts Configuration 2.3//EN"
    "http://struts.apache.org/dtds/struts-2.3.dtd">
<struts>
… …
    <package name="struts-default" abstract="true">
        <interceptors>
 … …
        </interceptors>
        <!-- 默认拦截器 -->
        <default-interceptor-ref name="defaultStack">
    </package>
</struts>
```

Struts 2 框架系统默认包 struts-default 定义了一系列拦截器和拦截器栈，同时使用 <default-interceptor - ref/> 指定了一个默认拦截器 defaultStack。

在配置文件中，一旦为包指定了默认拦截器，该拦截器会对包内所有 action 起作用；如果包中的 action 显式指定了拦截器，那么默认拦截器就会失去作用。包 chapter06 中没有指定默认拦截器，但是该包继承了 struts-default，同时继承了该包的默认拦截器 defaultStack。所以 chapter06 包内所有 action 处理用户请求时都会使用默认拦截器 defaultStack，但是前提条件是该 action 中没有显式指定拦截器。如果该 action 显式指定了拦截器，那么想要使用系统默认拦截器，必须显式加入默认拦截器的引用。

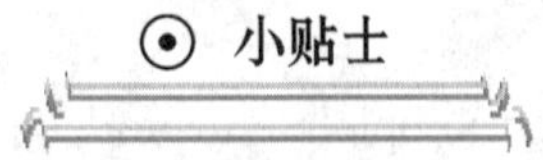

在指定包内默认拦截器时，只能使用一个 <default-interceptor-ref/> 元素，即每个包内只能定义一个默认拦截器。

标准的默认拦截器配置，如示例代码 6-13 所示：

示例代码 6-13 标准的默认拦截器配置代码

```
<!DOCTYPE struts PUBLIC
    "-//Apache Software Foundation//DTD Struts Configuration 2.3//EN"
    "http://struts.apache.org/dtds/struts-2.3.dtd">
<struts>
    <package name=" 包名 " >
        <interceptors>
            <interceptor name=" 拦截器 1" class=" 拦截器 1 实现类 "/>
            <interceptor name=" 拦截器 2" class=" 拦截器 2 实现类 "/>
            … …
            <interceptor-stack name=" 拦截器栈 1">
            … …
            </interceptor-stack>
            … …
        </interceptors >
        … …
        <default-interceptor-ref name=" 拦截器或者拦截器栈名称 "/>
        <action name="Action1" class="Action1 实现类 ">
        … …
        </action>
        … …
    </package>
</struts>
```

如果需要将多个拦截器都设置为一个包的默认拦截器，最好的方式就是将这些拦截器组成一个拦截器栈，再将该拦截器栈设置为该包的默认拦截器。这样使配置文件简单易懂，不易造成混乱。

提供默认拦截器的目的是避免为每个 action 配置相同的拦截器，使用默认拦截器，包内所有 action 都会自动使用默认拦截器，避免了配置代码的重复使用。同前例使用拦截器一样，在定义包默认拦截器时，可以初始化拦截器的参数，同样使用 <param/> 子元素。

配置拦截器的相关配置元素：

- <interceptors/>：定义拦截器，该元素可以包含拦截器定义子元素和拦截器栈定义子元素。使用 <interceptor/> 和 <interceptor-stack/> 定义拦截器和拦截器栈。
- <interceptor/>：定义拦截器，定义拦截器时需要指定两个属性，即 name 和 class。name 用于指定该拦截器名称；class 用于指定该拦截器的实现类。
- <interceptor-stack/>：定义拦截器栈，该元素中指定了多个 <interceptor-ref/> 元素，用于指定该拦截器栈所包含的拦截器或者拦截器栈名称。
- <interceptor-ref/>：该元素引用一个拦截器或者拦截器栈，只需要指定一个 name 属性，即前面已经定义好了的拦截器或者拦截器栈名称。该元素可以作为拦截器栈定义 <interceptor-stack/> 子元素，或者是 Action 中使用拦截器或拦截器栈引用。
- <param/>：该元素用于指定拦截器的参数，可以作为 <interceptor/> 元素的子元素。

6.3　自定义拦截器

作为“框架（framework）”，可扩展性是不可或缺的，因为世上没有放之四海而皆准的东西。虽然，Struts 2 为我们提供如此丰富的拦截器实现，但是这并不意味我们失去创建自定义拦截器的能力，恰恰相反，在 Struts 2 中自定义拦截器是相当容易的一件事。

> 大家在开始着手创建自定义拦截器前，切记以下原则：拦截器必须是无状态的，不要使用 API 提供的 ActionInvocation 之外的任何东西。

要求拦截器是无状态的原因是：Struts 2 不能保证为每一个请求或者 action 创建一个实例，所以，如果拦截器带有状态，会引发并发问题。所有的 Struts 2 的拦截器都直接或间接实现接口：com.opensymphony.xwork2.interceptor.Interceptor。除此之外，大家可能更喜欢继承类 com.opensymphony. xwork2.interceptor.AbstractInterceptor。

6.3.1　实现拦截器类

Struts 2 支持开发者定义自己的拦截器实现类，对系统提供的默认拦截器进行扩展和补充，开发一个自定义的拦截器实现类，在 Struts 2 框架内是一个非常简单的事情。

Struts 2 框架为开发自定义拦截器实现类提供了一个 Interceptor 接口，用户可以实现该接口来开发自定义拦截器实现类。Interceptor 接口定义，如示例代码 6-14 所示。

示例代码 6-14 Interceptor 接口定义

```
public interface Interceptor extends Serializable
{
        public void init();
        public void destroy();
        public String intercept(ActionInvocation invocation);
}
```

拦截器是无状态的，Struts 2 框架不会为每个处理用户请求的 action 建立一个实例。该接口中定义了三个方法。

（1）init()：在拦截器被初始化之后，执行拦截之前，系统会调用该方法。对于每个拦截器来说，该方法只会执行一次。所以，该方法主要用于处理一些一次性资源。

（2）destroy()：该方法跟 init() 方法相对应。在拦截器被销毁之前，系统将会调用该方法。一般情况下，该方法用于关闭 init() 方法打开的一些资源，例如数据库连接。

（3）String intercept(ActionInvocation invocation)：该方法是实现用户拦截的处理方法。该方法同 action 的 execute() 方法一样，返回 String 的结果，来对应配置文件中定义的逻辑视图名称。如果该方法直接返回一个字符串，框架将会跳转到该字符串对应的逻辑视图，不会调用被拦截的 action。invocation 参数包含了被拦截的 action 的引用，可以通过该参数的 invoke() 方法执行 action 的 execute() 方法或者执行下一个拦截器。

除此之外，继承类 com.opensymphony.xwork2.interceptor.AbstractInterceptor 是更简单的一种实现拦截器类的方式，因为此类提供了 init() 和 destroy() 方法的空实现，这样我们只需要实现 intercept() 方法。如示例代码 6-15 所示。

示例代码 6-15 自定义拦截器实现类 SimpleInterceptor

```
package com.xtgj.struts2.chapter06.intercepter;

import java.util.Date;

import com.opensymphony.xwork2.ActionInvocation;
import com.opensymphony.xwork2.interceptor.AbstractInterceptor;
import com.xtgj.struts2.chapter06.action.Reg;

public class SimpleInterceptor extends AbstractInterceptor {
    // 重写 intercept() 方法
    public String intercept(ActionInvocation arg0) throws Exception {
        // 获得被拦截的 Action 引用
```

```
        Reg reg = (Reg) arg0.getAction();
        System.out.println(" 拦截器信息：启动拦截器，拦截 Action 时间：" + new Date());
        // 执行 Action 或者执行下一个拦截器
        String result = arg0.invoke();
        // 提示 Action 执行完毕
        System.out.println(" 拦截器信息 ;Action 执行完毕时间：" + new Date());
        return result;
    }
}
```

用户自定义拦截器 SimpleInterceptor 继承了 AbstractInterceptor 类，并重写了 intercept() 方法，在该方法中，拦截器拦截 action 的执行，在拦截后，打印一条信息，显示拦截器已经拦截；然后调用 invoke() 方法，来执行 action 或者是下一个拦截器，返回 String 类型的结果。

用户在重写 intercept() 方法时，会获得 ActionInvocation 参数类型参数，通过该类型参数 arg0 来获得被拦截 action 的引用，即获得了 action 的控制权限。用户可以在拦截器中调用 action 中的 getter 和 setter 方法来获得用户请求参数，也可以对参数进行预先处理，即拦截器完全拦截了 action 的执行。同样，该拦截器可以不调用 invoke() 方法，即拦截用户请求不需要执行 action，直接从拦截器返回 result 结果。

6.3.2　使用自定义拦截器

开发者编写了自己的拦截器实现类后，就可以像系统提供的拦截器一样，在配置文件中定义和使用了。要使用自定义的拦截器，可以执行定义拦截器和使用拦截器两个步骤。

- 定义拦截器：使用 <interceptor/> 元素在包中定义该拦截器；
- 使用拦截器：使用 <interceptor-ref/> 元素在 action 中使用该拦截器。

以下就是定义、使用该拦截器的配置文件，如示例代码 6-16 所示。

示例代码 6-16　自定义拦截器的配置

```
<package name="chapter06" extends="struts-default">
        <interceptors>
            自定义拦截器
            <interceptor name="simpleInterceptor"
                class="com.xtgj.struts2.chapter06.intercepter.SimpleInterceptor" />
        </interceptors>
        <action name="Reg"
            class="com.xtgj.struts2.chapter06.action.Reg">
            <result name="success">/success.jsp</result>
            <result name="input">/reg.jsp</result>
             引用拦截器
```

```
            <interceptor-ref name="defaultStack" />
            <interceptor-ref name="simpleInterceptor" />
        </action>
</package>
```

6.3.3 自定义拦截器实例

以下案例演示通过继承 AbstractInterceptor，实现授权拦截器。

首先，创建授权拦截器类 com.xtgj.struts2.chapter06.intercepter.AuthorizationInterceptor，如示例代码 6-17 所示。

示例代码 6-17 自定义授权拦截器类 AuthorizationInterceptor

```
package com.xtgj.struts2.chapter06.intercepter;

import java.util.Map;
import com.opensymphony.xwork2.Action;
import com.opensymphony.xwork2.ActionInvocation;
import com.opensymphony.xwork2.interceptor.AbstractInterceptor;
public class AuthorizationInterceptor extends AbstractInterceptor {
    @Override
    public String intercept(ActionInvocation ai) throws Exception {
        Map session = ai.getInvocationContext().getSession();
        String role = (String) session.get("ROLE");
        if (null != role) {
            Object o = ai.getAction();
            if (o instanceof RoleAware) {
                RoleAware action = (RoleAware) o;
                action.setRole(role);
            }
            return ai.invoke();
        } else {
            return Action.LOGIN;
        }
    }
}
```

以上代码相当简单，我们通过检查 session 是否存在键为“ROLE”的字符串，判断用户是否登录。如果用户已经登录，将角色放到 action 中，调用 action；否则，拦截直接返回 Action.LOGIN 字段。为了方便将角色放入 action，可定义接口 com.xtgj.struts2.chapter06.intercepter.

RoleAware，如示例代码 6-18 所示。

示例代码 6-18　接口 RoleAware 的代码

```
package com.xtgj.struts2.chapter06.intercepter;
public interface RoleAware {
    void setRole(String role);
}
```

接着，创建 Action 类 com.xtgj.struts2.chapter06.action.AuthorizatedAccess 模拟访问受限资源，其作用就是通过实现 RoleAware 获取角色，并将其显示到 ShowUser.jsp 中，如示例代码 6-19 所示。

示例代码 6-19　AuthorizatedAccess Action 代码

```
package com.xtgj.struts2.chapter06.action;
import com.opensymphony.xwork2.ActionSupport;
import com.xtgj.struts2.chapter06.intercepter.RoleAware;
public class AuthorizatedAccess extends ActionSupport implements RoleAware {
    private String role;
    public void setRole(String role) {
        this.role = role;
    }
    public String getRole() {
        return role;
    }
    @Override
    public String execute() {
        return SUCCESS;
    }
}
```

ShowUser.jsp 页面代码如示例代码 6-20 所示。

示例代码 6-20　ShowUser.jsp 页面代码

```
<%@ page contentType="text/html;charset=UTF-8"%>
<%@taglib prefix="s" uri="/struts-tags"%>
<html>
    <head>
        <title>Authorizated User</title>
```

```
        </head>
        <body>
                <h1>
                        Your role is:
                        <s:property value="role" />
                </h1>
        </body>
</html>
```

然后，创建 com.xtgj.struts2.chapter06.action.Roles 初始化角色列表，如示例代码 6-21 所示。

示例代码 6-21　Roles.java 代码

```
package com.xtgj.struts2.chapter06.action;
import java.util.Hashtable;
import java.util.Map;
public class Roles {
        public Map<String, String> getRoles() {
                Map<String, String> roles = new Hashtable<String, String>(2);
                roles.put("EMPLOYEE", "Employee");
                roles.put("MANAGER", "Manager");
                return roles;
        }
}
```

接下来，新建 Login.jsp 实例化 com.xtgj.struts2.chapter06.action.Roles，并将其 roles 属性赋予 <s:radio> 标志，如示例代码 6-22 所示。

示例代码 6-22　Login.jsp 页面代码

```
<%@ page contentType="text/html;charset=UTF-8"%>
<%@taglib prefix="s" uri="/struts-tags"%>
<html>
        <head>
                <title>Login</title>
        </head>
        <body>
                <h1>
                        Login
                </h1>
                Please select a role below:
```

```
            <s:bean id="roles" name="com.xtgj.struts2.chapter06.action.Roles" />
            <s:form action="Login">
                <s:radio list="#roles.roles" value="'EMPLOYEE'" name="role"
                    label="Role" />
                <s:submit />
            </s:form>
        </body>
</html>
```

创建 action 类 com.xtgj.struts2.chapter06.action.Login 将 role 放到 session 中，并转到 action 类 com.xtgj.struts2.chapter06.action.AuthorizatedAccess，如示例代码 6-23 所示。

示例代码 6-23　Login Action 类代码

```
package com.xtgj.struts2.chapter06.action;
import java.util.Map;
import org.apache.struts2.interceptor.SessionAware;
import com.opensymphony.xwork2.ActionSupport;
public class Login extends ActionSupport implements SessionAware {
    private String role;
    private Map session;
    public String getRole() {
        return role;
    }
    public void setRole(String role) {
        this.role = role;
    }
    public void setSession(Map session) {
        this.session = session;
    }
    @Override
    public String execute() {
        session.put("ROLE", role);
        return SUCCESS;
    }
}
```

最后，配置 struts.xml 文件，如示例代码 6-24 所示。

示例代码 6-24 struts.xml 配置代码

```
<?xml version="1.0" encoding="UTF-8" ?>
<!DOCTYPE struts PUBLIC
    "-//Apache Software Foundation//DTD Struts Configuration 2.3//EN"
    "http://struts.apachc.org/dtds/struts-2.3.dtd">
<struts>
    <include file="struts-default.xml" />
    <package name="InterceptorDemo" extends="struts-default">
        <interceptors>
            <interceptor name="auth"
                class="com.xtgj.struts2.chapter06.intercepter.AuthorizationInterceptor" />
        </interceptors>
        <action name="Login" class="com.xtgj.struts2.chapter06.action.Login">
            <result type="chain">AuthorizatedAccess</result>
        </action>
        <action name="AuthorizatedAccess"
            class="com.xtgj.struts2.chapter06.action.AuthorizatedAccess">
            <result name="login">/Login.jsp</result>
            <result name="success">/ShowUser.jsp</result>
            <interceptor-ref name="auth" />
        </action>
    </package>
</struts>
```

发布运行应用程序，在浏览器地址栏中输入："http://localhost:8080/Struts2_Chapter06/Authorizated Access.action"。由于此时，session 还没有键为"ROLE"的值，所以返回 Login.jsp 页面，如图 6-4 所示。

图 6-4 Login.jsp 界面

选中 Employee，点击 Submit，出现图 6-5 所示页面。

图 6-5　ShowRole.jsp 界面

6.4　小结

- ✓ 拦截器是 Struts 2 比较重要的一个功能。
- ✓ 通过正确地使用拦截器，我们可以编写可复用性的代码。
- ✓ 拦截器是 Struts 2 的基石。

6.5　英语角

employee　　雇员
roles　　角色
interceptor　　拦截器

6.6　作业

1. 简述 Struts 2 拦截器的意义。
2. 用一个实例说明 Struts 2 拦截器的工作原理。
3. 简述 Struts 2 拦截器的应用。

6.7 思考题

认真体会默认拦截器和自定义拦截器的不同，了解在实际应用中的注意事项。

6.8 学员回顾内容

1. 拦截器的配置方法。
2. 自定义拦截器的实现。

第 7 章　Struts 2 标签库

学习目标

✧ 了解标签的主要特性。
✧ 理解标签的组成结构。
✧ 掌握标签库的使用。

课前准备

✧ Struts 2 标签库的组成。
✧ Struts 2 标签库的分类。

本章简介

本章主要介绍 Struts 2 的标签库的组成以及如何使用，学习控制标签以及数据标签的使用，了解标签的主要特性。

7.1　Struts 2 标签库概述

Struts 2 框架的标签库简化了页面输出，并支持更加复杂而丰富的功能，相对 Struts 1.x 有了巨大的进步。

7.1.1　标签库简介

JSP 标签库可以被看成是一种通过 JavaBean 生成基于 XML 脚本的方法。从概念上讲，标签就是很简单而且可重用的代码结构。

7.1.2　Struts 2 标签库组成

Struts 2 框架的标签库可以分为以下 3 类。

- 用户界面标签（UI 标签）：主要用来生成 HTML 元素的标签；
- 非用户界面标签：主要用于数据访问、逻辑控制等；
- Ajax 标签：Ajax 是 Web 2.0 阶段系列技术和相关产品服务中非常重要的一种技术，其全称是异步 JavaScript 和 XML（即 Asynchronous JavaScript and XML）。该标签用来支持 Ajax 技术。

其中，用户界面标签（UI 标签），也可分为下面两类。

（1）表单标签：表单标签主要用于生成 HTML 页面的 form 元素，以及普通表单元素的标签。

（2）非表单标签：非表单标签主要用于生成页面上的 tree、Tab 页等。

非界面标签，即非 UI 标签，也可以分为以下两类。

（1）数据访问标签：主要包含用于输出值栈（ValueStack）中的值、完成国际化等功能的标签。

（2）流程控制标签：主要包含用于实现分支、循环等流程控制的标签。

Struts 2 框架的标签库分类如图 7-1 所示。

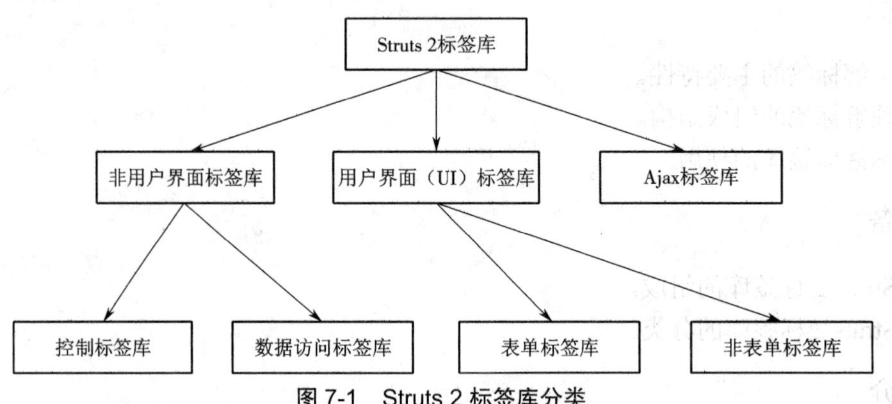

图 7-1 Struts 2 标签库分类

7.1.3 Struts 2 标签库的使用

前文介绍了标签库的组成，接着我们了解一下开发标签库的主要步骤。开发标签库有两个主要的步骤：开发标签库实现类和定义标签库文件。在 Struts 2 框架中，提供了标签的处理实现类和标签库定义文件。

读者可以在 struts2-core-2.3.15.3.jar 压缩文件的 META-INF 目录下找到 struts-tags.tld 文件，如图 7-2 所示。

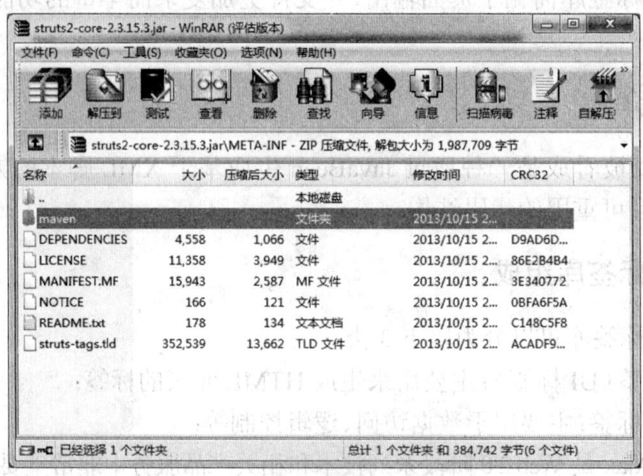

图 7-2 Struts 2 的标签库定义文件

struts-tags.tld 是一个标准的 XML 文件，其内容片段如示例代码 7-1 所示。

示例代码 7-1　struts-tags.tld 内容片段

```
<?xml version="1.0" encoding="UTF-8"?>
<taglib xmlns="http://java.sun.com/xml/ns/j2ee" xmlns:xsi="http://www.w3.org/2001/
XMLSchema-instance" version="2.0" xsi:schemaLocation="http://java.sun.com/xml/ns/j2ee
http://java.sun.com/xml/ns/j2ee/web-jsptaglibrary_2_0.xsd">
    <tlib-version>2.2.3</tlib-version>
    <-- 指定标签库默认的名称 -->
    <short-name>s</short-name>
    <-- 指定标签库默认的 URI -->
    <uri>/struts-tags</uri>
    <display-name>"Struts Tags"</display-name>
    <description><![CDATA["To make it easier to access dynamic data;
                    the Apache Struts framework includes a library of custom tags.
                    The tags interact with the framework's validation and interna-
tionalization features;
                    to ensure that input is correct and output is localized.
                    The Struts Tags can be used with JSP FreeMarker or Veloci-
ty."]]></description>
    <tag>
      <name>action</name>
      <tag-class>org.apache.struts2.views.jsp.ActionTag</tag-class>
      <body-content>JSP</body-content>
      <description><![CDATA[Execute an action from within a view]]></description>
      <attribute>
        <name>executeResult</name>
        <required>false</required>
        <rtexprvalue>false</rtexprvalue>
        <description><![CDATA[Whether the result of this action (probably a view)
should be executed/rendered]]></description>
      </attribute>
      <attribute>
        <name>flush</name>
        <required>false</required>
        <rtexprvalue>false</rtexprvalue>
        <description><![CDATA[Whether the writer should be flush upon end of action
component tag, default to true]]></description>
```

```
            </attribute>
            <attribute>
                <name>id</name>
                <required>false</required>
                <rtexprvalue>false</rtexprvalue>
                <description><![CDATA[id for referencing element. For UI and form tags it will
be used as HTML id attribute]]></description>
            </attribute>
            ......
    </tag>
    ......
```

如果需要在 JSP 页面中使用标签库，则需要使用 <taglib-location/> 元素指定标签库定义文件（TLD）文件的位置。一个标准的导入 Struts 2 标签库的指令格式如示例代码 7-2 所示。

示例代码 7-2 JSP 使用 Struts 2 标签库的格式

```
<%@ page language="java" pageEncoding="UTF-8"%>
<!-- 导入 Struts 2 标签库 -->
<%@ taglib uri="/struts-tags" prefix="s" %>
```

使用上述代码导入 Struts 2 框架的系统标签库，其中 s 为标签库默认的前缀，在 JSP 中引用标签。接下来，我们开始分类介绍 Struts 2 中的标签。

7.2 控制标签

控制标签主要用于完成流程控制，例如分支、循环等操作。控制标签包含以下几类。

- if：用于控制选择输出的标签；
- elseif：同 if 标签结合使用，用来控制选择输出；
- else：同 if 标签结合使用，用来控制选择输出；
- append：用来将多个集合拼接为一个新的集合；
- generator：为一个字符串解析器，用来将一个字符串解析为一个集合；
- iterator：迭代器，用来迭代输出集合数据；
- merge：用来将多个集合拼接为一个新的集合，同 append 有所区别。

7.2.1 if/elseif/else 标签

if/ elseif/ else 标签用来控制流程分支，同 Java 的流程控制相似，都是用于根据一个 boolean 表达式的值来决定是否进行相关的表达式操作。if/ elseif 标签属性 test 为必填属性，是一个 boolean 类型值，决定是否显示 if 标签内容。该类控制标签标准格式如示例代码 7-3 所示。

示例代码 7-3　if/ elseif/ else 标签使用格式

```
<s:if test=" 表达式 ">
    ……
</s:if>
<s:elseif test=" 表达式 ">
    ……
</s:elseif>
<s:else>
    ……
</s:else>
```

下面用一个示例来演示该标签，如代码 7-4 所示。

示例代码 7-4　if/ elseif/ else 标签使用示例

```
<%@ page language="java" pageEncoding="UTF-8"%>
<%@ taglib prefix="s" uri="/struts-tags"%>
<html>
    <head>
        <title>Struts 2 控制标签示例 </title>
    </head>
    <body>
        <!-- 定义一个 testname 属性 -->
        <s:set name="testname" value="%{'Java'}" />
        <!-- 使用 if 标签判断 -->
        <s:if test="%{#testname=='Java'}">
            <div>
                <s:property value="%{#testname}" />
            </div>
        </s:if>
        <s:elseif test="%{#testname=='Jav'}">
            <div>
                <s:property value="%{#testname}" />
            </div>
        </s:elseif>
        <s:else>
            <div>
                testname 不是“Java”
```

```
                </div>
            </s:else>
        </body>
    </html>
```

本示例中使用了 OGNL 表达式，请参考上机部分。运行结果如图 7-3 所示。

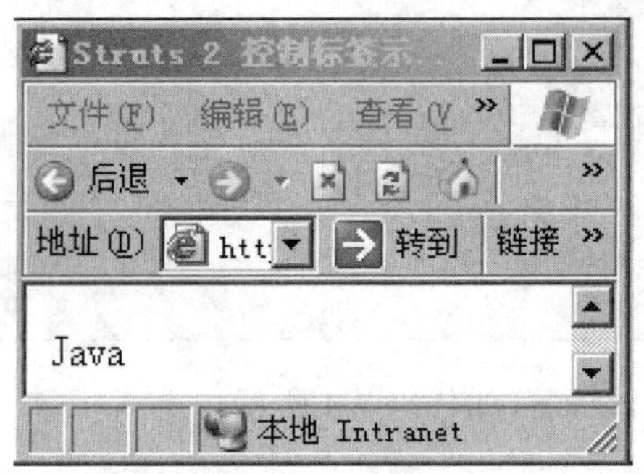

图 7-3 运行结果

示例中，先使用 <s:set /> 标签定义一个 testname 属性，并赋值为“Java”，接下来使用 <s:if/> 来判断该属性值是否为“Java”，如果是，则显示该属性值。

7.2.2 iterator 标签

iterator 标签主要用来对集合属性的迭代，其中集合属性类型可以是 list、map 或者是数组。使用 <s:iterator/> 进行迭代输出时，该标签属性如下。

- id：指定集合元素的 ID；
- value：可选属性，value 指定被迭代输出的集合属性，被迭代的集合通常都是使用 OGNL 表达式来指定。如果没有指定 value 属性，则使用值栈（ValueStack）栈顶的集合；
- status：可选属性，为一个 boolean 类型值，该属性指定了迭代时的 IteratorStatus 实例，通过该实例可以判断当前迭代元素的属性值，例如是否位于最后一个元素，或者是当前迭代元素索引值。默认值为 FALSE。

IteratorStatus 实例包含以下几个方法。

- int getCount()：返回当前迭代过元素的总数；
- int getIndex()：返回当前迭代元素的索引；
- boolean isEven()：判断当前迭代元素是否为偶数；
- boolean isOdd()：判断当前迭代元素是否为奇数；
- boolean isFirst()：判断当前迭代元素是否为第一个元素；
- boolean isLast()：判断当前迭代元素是否为最后一个元素。

下面用一个示例来演示如何使用 iterator 标签。

首先，定义一个业务控制器 action 实现类，类中定义了一个 list 属性和一个 map 属性，如示例代码 7-5 所示。

示例代码 7-5　IteratorTag Action 代码

```
package com.xtgj.struts2.chapter07.action;
// 省略 import
public class IteratorTag extends ActionSupport {
        private List myList;
        private Map myMap;
        public String execute() throws Exception {
                myList = new ArrayList();
                myList.add(" 第一个元素 ");
                myList.add(" 第二个元素 ");
                myList.add(" 第三个元素 ");
                myMap = new HashMap();
                myMap.put("key1", " 第一个元素 ");
                myMap.put("key2", " 第二个元素 ");
                myMap.put("key3", " 第三个元素 ");
                return SUCCESS;
        }
        public List getMyList() {
                return myList;
        }
        public void setMyList(List myList) {
                this.myList = myList;
        }
        public Map getMyMap() {
                return myMap;
        }
        public void setMyMap(Map myMap) {
                this.myMap = myMap;
        }
}
```

接着，在 struts.xml 文件中加入配置内容，如示例代码 7-6 所示。

示例代码 7-6　IteratorTag Action 配置

```
<action name="iteratorTag"
                class="com.xtgj.struts2.chapter07.action.IteratorTag">
                <result name="success">/iteratorTag.jsp</result>
</action>
```

然后，创建 JSP 视图资源，如示例代码 7-7 所示。

示例代码 7-7　iteratorTag.jsp 代码

```
<%@ page language="java" pageEncoding="UTF-8"%>
<%@ taglib prefix="s" uri="/struts-tags"%>
<html>
    <head>
        <title>Iterator 标签示例！ </title>
    </head>
    <body>
        <h1>
            <span style="background-color: #FFFFcc">Iterator 标签示例 </span>
        </h1>
        <h2>
                显示 List
        </h2>
        <table>
            <s:iterator value="{' 第一个元素 ',' 第二个元素 '}" status="st">
                <tr>
                    <td><s:property value="#st.getIndex()" /></td>
                    <td><s:property /></td>
                </tr>
            </s:iterator>
        </table>
        <h2>
            显示 List 属性
        </h2>
        <table>
            <s:iterator value="myList" status="st">
                <tr>
                    <td><s:property value="#st.getIndex()" /></td>
                    <td><s:property /></td>
                </tr>
            </s:iterator>
        </table>

        <h2>
        显示 Map
```

```
            </h2>
            <table>
                    <s:iterator  value="#{'key1':' 第一个元素 ','key2':' 第二个元素 '}"
status="st">
                        <tr>
                            <td><s:property value="#st.getIndex()" /></td>
                            <td><s:property /></td>
                        </tr>
                  </s:iterator>
            </table>
            <h2>
                        显示 Map 属性
            </h2>

            <table>
                  <s:iterator value="myMap" status="st">
                        <tr>
                            <td><s:property value="#st.getIndex()" /></td>
                            <td><s:property /></td>
                        </tr>
                  </s:iterator>
            </table>
      </body>
  </html>
```

最后发布运行程序，在浏览器中输入 http://localhost:8080/Struts2_Chapter07/iteratorTag.action，运行界面如图 7-4 所示。

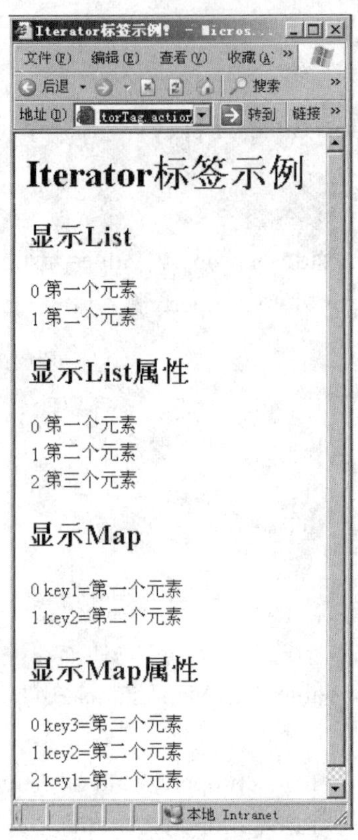

图 7-4 运行结果

7.2.3 append 标签

append 标签用于将多个集合对象拼接在一起，组成一个新的集合，这样拼接的目的是可以将多个集合使用一个 <iterator/> 标签完成迭代。标签属性 id: 指定了集合元素的 ID。

append 标签中可以使用 param 来指定用来拼接的子集合，可以指定多个子集合， append 标签会将指定的一个或者多个集合拼接为一个集合。以下是一个使用 append 标签的例子。

首先，创建 AppendTag 类，在该类中建立两个 list 属性和两个 map 属性，如示例代码 7-8 所示。

示例代码 7-8 AppendTag 代码

```
package com.xtgj.struts2.chapter07.action;
// 省略 import
public class AppendTag extends ActionSupport {
    private List myList1, myList2;
    private Map myMap1, myMap2;
    public String execute() throws Exception {
        myList1 = new ArrayList();
        myList2 = new ArrayList();
```

```
        myList1.add(" 第一个集合 # 第一个元素 ");
        myList1.add(" 第一个集合 # 第二个元素 ");
        myList1.add(" 第一个集合 # 第三个元素 ");
        myList2.add(" 第二个集合 # 第一个元素 ");
        myList2.add(" 第二个集合 # 第二个元素 ");
        myList2.add(" 第二个集合 # 第三个元素 ");
        myMap1 = new HashMap();
        myMap2 = new HashMap();
        myMap1.put("key1", " 第一个集合 # 第一个元素 ");
        myMap1.put("key2", " 第一个集合 # 第二个元素 ");
        myMap1.put("key3", " 第一个集合 # 第三个元素 ");
        myMap2.put("key1", " 第二个集合 # 第一个元素 ");
        myMap2.put("key2", " 第二个集合 # 第二个元素 ");
        myMap2.put("key3", " 第二个集合 # 第三个元素 ");
        return SUCCESS;
    }
    public List getMyList1() {
        return myList1;
    }
    public void setMyList1(List myList1) {
        this.myList1 = myList1;
    }
    public List getMyList2() {
        return myList2;
    }
    public void setMyList2(List myList2) {
        this.myList2 = myList2;
    }
    public Map getMyMap1() {
        return myMap1;
    }
    public void setMyMap1(Map myMap1) {
        this.myMap1 = myMap1;
    }
    public Map getMyMap2() {
        return myMap2;
    }
    public void setMyMap2(Map myMap2) {
```

```
            this.myMap2 = myMap2;
        }
}
```

接着，创建 JSP 视图 appendTag.jsp，使用 append 标签，如示例代码 7-9 所示。

示例代码 7-9　appendTag.jsp 代码

```
<%@ page language="java" pageEncoding="UTF-8"%>
<%@ taglib prefix="s" uri="/struts-tags"%>
<html>
    <head>
        <title>append 标签示例！</title>
    </head>
    <body>
        <h1>
            <span style="background-color: #FFFFCC">Append 标签示例 </span>
        </h1>
        <h2>
            拼接 List 属性
        </h2>
        <table>
            <s:append id="newList">
                <s:param value="myList1" />
                <s:param value="myList2" />
            </s:append>
            <s:iterator value="#newList" id="name" status="st">
                <tr>
                    <td><s:property value="#st.getIndex()" /></td>
                    <td><s:property /></td>
                </tr>
            </s:iterator>
        </table>
        <h2>
            拼接 Map 属性
        </h2>
        <table>
            <s:append id="newMap">
```

```
                    <s:param value="myMap1" />
                    <s:param value="myMap2" />
                </s:append>
                <s:iterator value="#newMap" status="st">
                    <tr>
                        <td><s:property value="#st.getIndex()" /></td>
                        <td><s:property /></td>
                    </tr>
                </s:iterator>
            </table>
        </body>
    </html>
```

最后发布运行程序，在浏览器中输入 http://localhost:8080/Struts2_Chapter07/appendTag.action，运行界面如图 7-5 所示。

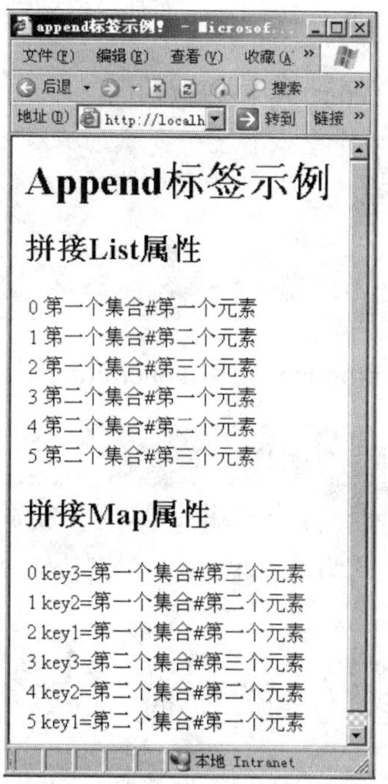

图 7-5　运行结果

7.2.4　generator 标签

generator 标签用来将指定的字符串按照规定的分隔符分解为多个子字符串，生成的多个子字符串可以使用 iterator 标签输出，该标签的属性如下。

- id：指定了集合元素的 ID；
- count：可选属性，为一个 Integer 类型值，指定生成集合中元素的总数；

- separator：必填属性，为一个 String 类型值，指定用来分解字符串的分隔符；
- val: 必填属性，为一个 String 类型值，指定被分解的字符串；
- converter：可选属性，为一个 Converter 类型实例，指定一个转换器，该转换器负责将集合中的每个字符串转换为对象。

以下是一个 generator 标签的实例，首先，创建 GeneratorAction 类，如示例代码 7-10 所示。

示例代码 7-10 GeneratorAction 代码

```
package com.xtgj.struts2.chapter07.action;
import com.opensymphony.xwork2.ActionSupport;
public class GeneratorAction extends ActionSupport {
	private String msg;
	public String getMsg() {
		return msg;
	}
	public void setMsg(String msg) {
		this.msg = msg;
	}
	public String execute() throws Exception {
		setMsg(" 第一个元素；第二个元素；第三个元素 ");
		return SUCCESS;
	}
}
```

其次，创建 JSP 视图 generatorTag.jsp，使用 generator 标签，如示例代码 7-11 所示。

示例代码 7-11 generatorTag.jsp 代码

```
<%@ page language="java" pageEncoding="UTF-8"%>
<%@ taglib prefix="s" uri="/struts-tags"%>
<html>
	<head>
		<title>Generator 标签示例！ </title>
	</head>
	<body>
		<h1>
			<span style="background-color: #FFFFcc">Generator 标签示例 </span>
		</h1>
		<table>
			<s:generator separator=";" val="msg" id="temp" count="3"></s:generator>
			<s:iterator  status="st"  value="#attr.temp">
```

```
                        <tr>
                            <td><s:property value="#st.getIndex()" /></td>
                            <td><s:property /></td>
                        </tr>
                    </s:iterator>
                </table>
            </body>
</html>
```

最后发布运行程序，在地址栏中输入“http://localhost:8080/Struts2_Chapter07/generatorTag.action”，运行结果如图 7-6 所示。如果在 generator 标签中指定了 count 属性，则该集合中最多有 count 个元素，多余的元素将会被忽略。

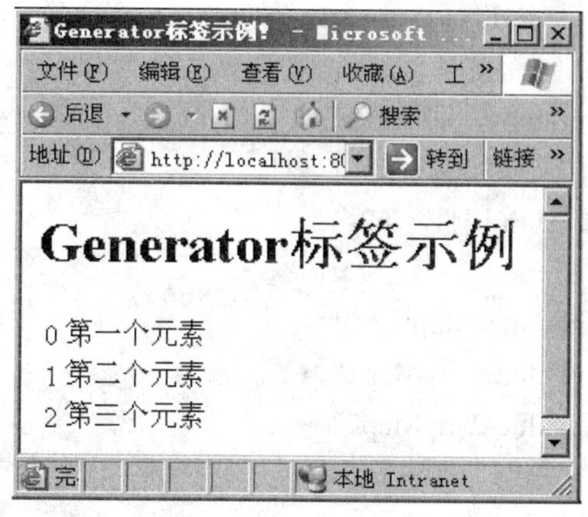

图 7-6　运行结果

7.2.5　merge 标签

merge 标签用于将多个集合拼接为一个集合，同 append 标签类似。但是同 append 标签不同的是在拼接集合时，新集合中集合元素的排列顺序不同。修改 append 标签代码的 JSP 视图，将 append 替换为 merge 标签，如示例代码 7-12 所示。

示例代码 7-12　mergeTag.jsp 代码

```
<%@ page language="java" pageEncoding="UTF-8"%>
<%@ taglib prefix="s" uri="/struts-tags"%>
<html>
    <head>
        <title>Merge 标签示例！</title>
    </head>
    <body>
```

```
        <h1>
            <span style="background-color: #FFFFcc">Merge 标签示例 </span>
        </h1>
        <h2> 拼接 List 属性 </h2>
        <table>
        <s:merge id="newList">
         <s:param value="myList1"/>
         <s:param value="myList2"/>
        </s:merge>
            <s:iterator value="#newList" id="name" status="st">
                <tr>
                    <td><s:property value="#st.getIndex()" /></td>
                    <td><s:property /></td>
            </tr>
            </s:iterator>
        </table>
        <h2> 拼接 Map 属性 </h2>
        <table>
        <s:merge id="newMap">
         <s:param value="myMap1"/>
         <s:param value="myMap2"/>
        </s:merge>
            <s:iterator value="#newMap"  status="st">
                <tr>
                    <td><s:property value="#st.getIndex()" /></td>
                    <td><s:property /></td>
                </tr>
            </s:iterator>

        </table>
    </body>
</html>
```

运行界面如图 7-7 所示。

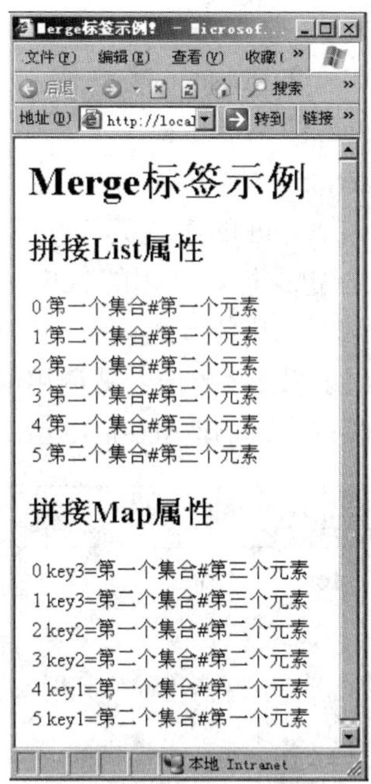

图 7-7　运行结果

7.3　数据标签

数据标签（Data Tags）主要用来提供数据访问功能，包含如下标签。

- action: 该标签用来直接调用一个 action，根据 executeResult 参数，可以将该 action 的处理结果包含到页面中；
- bean：用来创建一个 JavaBean 实例；
- date：用来格式化输出一个日期属性；
- i18n：用来指定国际化资源文件的 baseName；
- debug：用来生成一个调试链接，当单击该链接时，可以看到当前值栈中的内容；
- include：用来包含其他的页面资源；
- param：用来设置参数；
- property: 用来输出某个值，可以输出值栈，Stack Context 和 Action Context 中的值；
- push：用来将某个值放入值栈；
- set：用来设置一个新的变量；
- url：用来生成一个 URL 地址。

7.3.1　action 标签

action 标签允许开发者在 JSP 界面中直接调用 action，这种调用一般是通过指定 action 名称和命名空间来实现的。根据 executeResult 参数，可以将 action 的处理结果界面包含到当前页面中。该标签属性如下。

- executeResult：可选属性，为一个 Boolean 类型值，用来指定是否显示 action 的执行结果（通常为一个视图），默认值为 false，即不显示；
- id: 可选属性，用来引用该 action 的标识；
- name：必填属性，用来指定该 action 的实现类位置；
- namespace：可选属性，用来指定该标签调用的 action 所在的命名空间；
- ignoreContextParams：可选属性，用来指定该页面中的请求参数是否需要传入调用的 action。默认值为 false，即将本页面的请求参数传入被调用的 action。

以下是一个 action 标签的示例，首先，创建 ActionTagAction，如示例代码 7-13 所示。

示例代码 7-13　ActionTagAction 代码

```
package com.xtgj.struts2.chapter07.action;
// 省略 import
public class ActionTagAction extends ActionSupport {
      private String msg;
      public String execute() throws Exception {
            return SUCCESS;
      }
      public String doDefault() throws Exception {
            ServletActionContext.getRequest().setAttribute("ActionString",
                        "doDefault() 方法产生的字符串 ");
            return SUCCESS;
      }
      public String getMsg() {
            return msg;
      }
     public void setMsg(String msg) {
            this.msg = msg;
      }
}
```

然后，在配置文件中增加配置，如示例代码 7-14 所示。

示例代码 7-14　ActionTagAction 配置

```
<action name="actionTagAction1"
            class="com.xtgj.struts2.chapter07.action.ActionTagAction">
            <result name="success">/success.jsp</result>
</action>
<action name="actionTagAction2"
```

```
            class="com.xtgj.struts2.chapter07.action.ActionTagAction" method="default">
            <result name="success">/success.jsp</result>
</action>
```

接着，创建 success.jsp 页面，它是一个简单的 JSP 视图，如示例代码 7-15 所示。

示例代码 7-15　success.jsp 页面

```
<%@ page language="java" pageEncoding="UTF-8"%>
<%@ taglib prefix="s" uri="/struts-tags"%>
<html>
    <h2>Action 返回视图 </h2>
    <s:property value="msg"/>
    <s:property value="#attr.ActionString"/>
</html>
```

最后，创建视图 actionTag.jsp，该视图只是在页面上输出两个字符串信息，如示例代码 7-16 所示。

示例代码 7-16　actionTag.jsp 代码

```
<%@ page language="java" pageEncoding="UTF-8"%>
<%@ taglib prefix="s" uri="/struts-tags"%>
<html>
    <head>
        <title>Struts 2 控制标签示例 </title>
    </head>
    <body>
        <div>
            下面的 action 标签将会显示结果返回界面
        </div>
        <br/>
    <s:action name="actionTagAction1" executeResult="true" />
        <br/>
        <div>
            下面的 action 标签将会显示 dodefault 方法的结果界面
        </div>
        <br/>
        <s:action name="actionTagAction2" executeResult="true"
            ignoreContextParams="true" />
        <br/>
```

```
            <div>
                使用 action 标签，但是 executeResult 为 false，则不会显示结果界面
            </div>
            <s:action name="actionTagAction2" executeResult="false" />
            <s:property value="#attr.ActionString" />
        </body>
    </html>
```

示例代码 7-16 中使用了三个 action 标签，其中第二个 action 标签指定了 ignoreContextParams 属性为 true，即忽略 JSP 请求参数，不会将参数传递给 action；第三个 action 标签指定了 executeResult 属性为 false，即不在本页面显示 action 的执行结果界面。运行结果如图 7-8 所示。

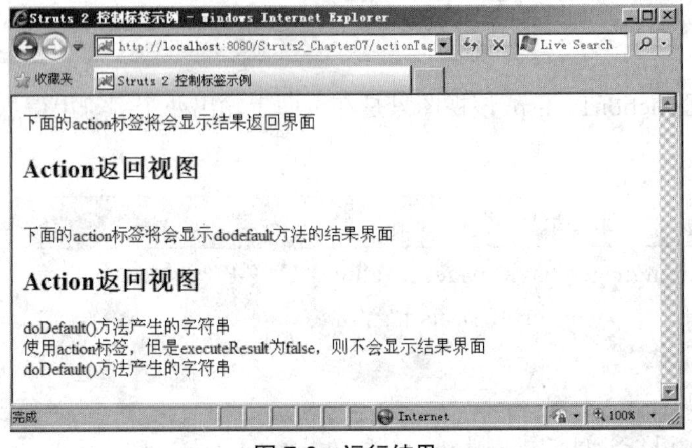

图 7-8 运行结果

7.3.2 bean 标签

bean 标签用来创建一个 JavaBean 的实例，以便在 JSP 视图中使用。该标签为 <s:bean/>，开发者可以在 bean 标签中使用 <s:param/> 标签来指定该 bean 中的属性值。实际上，使用 <s:param/> 标签就是调用了 action 中对应属性的 setter 方法。bean 标签有如下主要属性。

- id：可选属性，用来标识 JavaBean 的实例化对象；
- name：必填属性，该属性指定了要实例化的 JavaBean 实现类。

以下是一个 bean 标签的实例，首先，建立一个 User 类，如示例代码 7-17 所示。

示例代码 7-17 User 类代码

```
package com.xtgj.struts2.chapter07.action;
import java.sql.Date;
public class User {
    private String name;
```

```
        private int age;
        private Date birthday;
        public Date getBirthday() {
                return birthday;
        }
        public void setBirthday(Date birthday) {
                this.birthday = birthday;
        }
        public String getName() {
                return name;
        }
        public void setName(String name) {
                this.name = name;
        }
        public int getAge() {
                return age;
        }
        public void setAge(int age) {
                this.age = age;
        }
}
```

接着创建视图 beanTag.jsp，演示在 bean 标签内访问 JavaBean 实例和在 bean 标签外访问 JavaBean 实例，如示例代码 7-18 所示。

示例代码 7-18　beanTag.jsp 代码

```
<%@ page language="java" pageEncoding="UTF-8"%>
<%@ taglib prefix="s" uri="/struts-tags"%>
<html>
        <head>
                <title>Struts 2 控制标签示例 </title>
        </head>
        <body>
                <!-- 使用 bean 标签，并在标签内访问数据 -->
                <s:bean name="com.xtgj.struts2.chapter07.action.User">
                <s:param name="name" value="'pla'"/>
                <s:param name="age" value="30"/>
                <s:param name="birthday" value="'2007-01-01'"/>
```

```
            <!-- 由于在标签内,可以直接输出属性值 -->
            <s:property value="name" /><br>
            <s:property value="age" /><br>
            <s:property value="birthday" /><br>
            </s:bean>
            <!-- 使用 bean 标签,并在标签外访问数据,需要指定 id 属性 -->
            <s:bean name="com.xtgj.struts2.chapter07.action.User" id="user">
            <s:param name="name" value="'pla'"/>
            <s:param name="age" value="30"/>
            <s:param name="birthday" value="'2007-01-01'"/>
            </s:bean>
            <!-- 使用 id 属性,访问 StackContext 中的实例 -->
            <s:property value="#user.name" /><br>
            <s:property value="#user.age" /><br>
            <s:property value="#user.birthday" />
        </body>
    </html>
```

运行结果如图 7-9 所示。

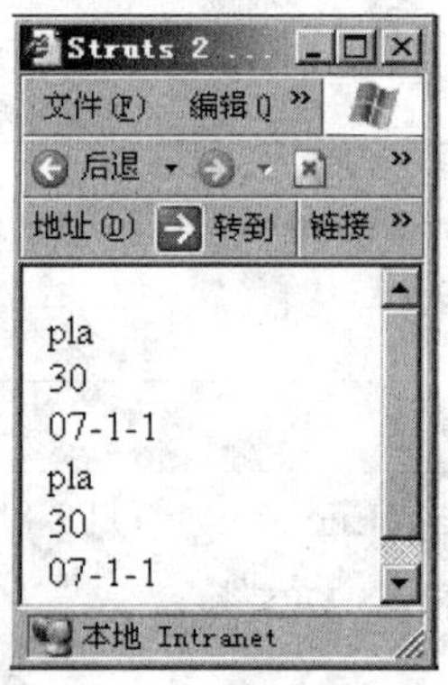

图 7-9 运行结果

7.3.3 date 标签

date 标签用来以快捷、简单的方式格式化输出一个日期值。开发者可以指定一个本地的日期格式,例如"DD/MM/YYYY hh:mm"。可以产生非常易读的日期信息。该标签还支持按照预定的输出格式输出日期值,一般是在资源文件中指定"struts.date.format"来实现。该标签属性介绍如下。

- id:可选属性,用来指定引用该元素的 id 值;
- nice:可选属性,为一个 Boolean 类型值,用于指定是否输出指定日期和当前日期之间的时差;默认为 false,即不输出;

- name：必选属性，用来指定要格式化输出的日期值；
- format：可选属性，如果指定了该属性，则会按照其规定的格式输出日期值。

首先，建立一个使用 date 标签示例的业务控制器，如示例代码 7-19 所示。

示例代码 7-19　DateTag Action 代码

```
package com.xtgj.struts2.chapter07.action;
// 省略 import
public class DateTag extends ActionSupport {
        private Date currentDate;
        public String execute() throws Exception {
                setCurrentDate(new Date());
                return SUCCESS;
        }
        public void setCurrentDate(Date date) {
                this.currentDate = date;
        }
        public Date getCurrentDate() {
                return currentDate;
        }
}
```

然后，在配置文件中加入 action 配置，如示例代码 7-20 所示。

示例代码 7-20　DateTag Action 配置

```
<action name="dateTag" class="com.xtgj.struts2.chapter07.action.DateTag">
                <result name="success">/dateTag.jsp</result>
</action>
```

最后，创建使用 date 标签的 JSP 视图 dateTag .jsp，如示例代码 7-21 所示。

示例代码 7-21　dateTag .jsp 代码

```
<%@ page language="java" pageEncoding="UTF-8"%>
<%@ taglib prefix="s" uri="/struts-tags"%>
<html>
  <head>
    <title>Date 标签示例 !</title>
  </head>
  <body>
    <h1><font color="#000080"> 日期输出格式 </font></h1>
```

```
<table border="1" width="35%" bgcolor="ffffcc">
    <tr>
        <td width="50%"><b><font color="#000080"> 日期格式 </font></b></td>
        <td width="50%"><b><font color="#000080"> 日期 </font></b></td>
    </tr>
    <tr>
        <td width="50%">Day/Month/Year</td>
        <td width="50%"><s:date name="currentDate" format="dd/MM/yyyy" /></td>
    </tr>
    <tr>
        <td width="50%">Month/Day/Year</td>
        <td width="50%"><s:date name="currentDate" format="MM/dd/yyyy" /></td>
    </tr>
    <tr>
        <td width="50%">Month/Day/Year</td>
        <td width="50%"><s:date name="currentDate" format="MM/dd/yy" /></td>
    </tr>
    <tr>
        <td width="50%"> 年 / 月 / 日 </td>
        <td width="50%"><s:date name="currentDate" format="yy 年 MM 月 dd
日 " /></td>
    </tr>
    <tr>
        <td width="50%">Month/Day/Year Hour<B>:</B>Minute</td>
        <td width="50%"><s:date name="currentDate" format="DD/MM/YYYY. hh:
mm" /></td>
    </tr>
    <tr>
        <td width="50%">Month/Day/Year Hour<B>:</B>Minute<B>:</B>Sec-
ond</td>
        <td width="50%"><s:date name="currentDate" format="MM/dd/yy hh:m-
m:ss" /></td>
    </tr>
    <tr>
        <td width="50%">Nice Date (Current Date & Time)</td>
        <td width="50%"><s:date name="currentDate" nice="false" /></td>
    </tr>
    <tr>
```

```
                <td width="50%">Nice Date</td>
                <td width="50%"><s:date name="currentDate" nice="true" /></td>
            </tr>
        </table>
    </body>
</html>
```

运行结果如图 7-10 所示。

图 7-10　运行结果

Struts 2 中还支持很多功能强大的标签，在这里就不一一介绍了，请读者参考 API 实验。

7.4　小结

✓ 标签的主要特性。
✓ 标签的组成结构。

7.5 英语角

asynchronous 异步
current 当前
generator 发电机

7.6 作业

回顾标签的使用。

7.7 思考题

常用的控制标签和数据标签主要有什么？

7.8 学员回顾内容

1. 掌握常用的控制标签。
2. 掌握常用的数据标签。

第 8 章　Struts 2 国际化

学习目标

✧ 了解国际化的概念。
✧ 理解国际化的作用。
✧ 掌握国际化的使用。

课前准备

✧ 国际化资源文件的方式。
✧ 国际化具体的应用。

本章简介

本章主要介绍 Struts 2 的国际化的概念、国际化的作用，学习如何利用 Struts 2 实现国际化，了解国际化的实现过程。

8.1　国际化简介

国际化英文单词为 internationalization，该单词比较长，不方便记录，由于在 i 和 n 之间有 18 个字母，所以也称 i18n，这样使用起来比较简短方便。

国际化软件是指软件能够运行于不同的区域和语言环境中，根据用户所处的区域和语言来修饰软件界面和显示信息，使用户能够通过熟悉的界面环境来使用该软件。

8.2　Struts 2 的国际化支持

Struts 2 框架的底层国际化与 Java 国际化是一致的，作为一个良好的 MVC 框架， Struts 2 将 Java 的国际化功能进行了封装和简化，开发者使用起来会更加简单快捷。在 Struts 2 中需要国际化的有：JSP 页面的国际化、action 错误信息的国际化、转换错误信息的国际化、校验错误信息的国际化。

8.2.1 配置全局资源文件及输出国际化信息

下面先看一个例子——Hello World。该案例将演示如何根据用户浏览器的语言环境设置输出相应环境的 Hello World。基本步骤如下。

（1）准备资源文件，资源文件的命名格式如下：

```
baseName_language_country.properties
baseName_language.properties
baseName.properties
```

其中 baseName 是资源文件的基本名，我们可以自定义，但 language 和 country 必须是 Java 支持的语言和国家，如：

```
中国大陆：baseName_zh_CN.properties
美国：baseName_en_US.properties
```

（2）在 struts.xml 中通过 struts.custom.i18n.resources 常量把资源文件定义为全局资源文件，如：

```
<constant name="struts.custom.i18n.resources" value="globalMessages" />
```

- 上述常量配置中“globalMessages”为资源文件的基本名。
- Struts 2 有两个配置文件，struts.xml 和 struts.properties 都放在 WEB-INF/classes/ 下。
  - ✧ struts.xml 用于应用程序相关的配置；
  - ✧ struts.properties 用于 Struts 2 运行时（Runtime）的配置。

（3）在 src 文件夹中加入 globalMessages_en_US.properties 文件，内容如下：

```
HelloWorld=Hello World!
```

（4）在 src 文件夹中加入 globalMessages_zh_CN.properties 文件，内容如下：

```
HelloWorld= 你好，世界！
```

对于中文的属性文件，我们编写好后，应该使用 jdk 提供的 native2ascii 命令把文件转换为 unicode 编码的文件。命令的使用方式如下：

```
native2ascii 源文件 .properties 目标文件 .properties
```

（5）在 WebRoot 文件夹下加入 HelloWorld.jsp 文件，如示例代码 8-1 所示。

示例代码 8-1　HelloWorld.jsp 代码

```
<%@ page language="java" pageEncoding="UTF-8"%>
<%@taglib prefix="s" uri="/struts-tags"%>
<html>
    <head>
        <title>Hello World</title>
    </head>
    <body>
        <h2>
            <s:text name="HelloWorld" />
        </h2>
        <h2>
            <s:property value="%{getText('HelloWorld')}" />
        </h2>
    </body>
</html>
```

（6）发布运行应用程序，在浏览器地址栏中输入：http://localhost:8080/Struts2_Chapter08/HelloWorld.jsp，出现如图 8-1 所示的页面。

将浏览器的默认语言改为“英语（美国）”，刷新页面，出现如图 8-2 所示的页面。

图 8-1　中文输出

图 8-2　英文输出

以上例子的做法与 Struts 1.x 的做法相似，似乎并不能体现 Struts 2 的优势。不过，以上案例用了两种方法来显示国际化字符串，其输出是相同的。其实，这就是 Struts 2 的一个优势，因为它默认支持 EL，所示我们可以用 getText() 方法来简捷地取得国际化字符串。此外，更普遍的情况——在使用 UI 表单标志时，getText() 方法可以用来设置 label 属性，例如：

```
<s:textfield name="name" label="%{getText('UserName')}"/>
```

8.2.2 参数国际化字符串

许多情况下，我们都需要在运行时（runtime）为国际化字符插入一些参数，例如在输入验证提示信息的时候。在 Struts 2 中，我们通过以下两种方法可以做到这点。

（1）在资源文件的国际化字符串中使用 OGNL，格式为 ${ 表达式 }，例如：

```
validation.require=${getText(fileName)} is required
```

（2）使用 java.text.MessageFormat 中的字符串格式，格式为 { 参数序号（从 0 开始），格式类形（number | date | time | choice），格式样式 }，例如：

```
validation.between=Date must between {0, date, short} and {1, date, short}
```

在显示这些国际化字符时，同样有两种方法设置参数的值。

（1）使用标志的 value0、value1...valueN 的属性，如：

```
<s:text name="validation.required" value0="User Name"/>
```

（2）使用 param 子元素，这些 param 将按先后顺序，代入到国际化字符串的参数中，例如：

```
<s:text name="validation.required">
    <s:param value="User Name"/>
</s:text>
```

8.2.3 让用户方便地选择语言

开发国际化的应用程序时，有一个功能是必不可少的——让用户快捷地选择或切换语言。在 Struts 2 中，通过 ActionContext.getContext().setLocale(Locale arg) 可以设置用户的默认语言。不过，由于这是一个比较普遍的应用场景，所以 Struts 2 为用户提供了一个名为 i18n 的拦截器（Interceptor），并在默认情况下将其注册到拦截器链（Interceptor Chain）中。它的原理为在执行 action 方法前，i18n 拦截器查找请求中的一个名为“request_locale”的参数。如果其存在，拦截器就将其作为参数实例化 Locale 对象，并将其设为用户默认的区域（Locale）。最后，将此 Locale 对象保存在 session 的名为“WW_TRANS_I18N_LOCALE”的属性中。下面将提供一个完整示例演示它的使用方法。

新建 Locales.java，如示例代码 8-2 所示。

示例代码 8-2　Locales.java 代码

```
package com.xtgj.struts2.chapter08.locale;


import java.util.Hashtable;
import java.util.Locale;
```

```
import java.util.Map;

public class Locales {
    public Map<String, Locale> getLocales() {
        Map<String, Locale> locales = new Hashtable<String, Locale>(2);
        locales.put("American English", Locale.US);
        locales.put("Simplified Chinese", Locale.CHINA);
        return locales;
    }
}
```

创建 JSP 页面 LangSelector.jsp，如示例代码 8-3 所示。

示例代码 8-3　LangSelector.jsp 代码

```
<%@taglib prefix="s" uri="/struts-tags"%>
<script type="text/javascript">
<!--
    function langSelecter_onChanged() {
        document.langForm.submit();
    }
//-->
</script>
<s:set name="SESSION_LOCALE" value="#session['WW_TRANS_I18N_LOCALE']" />
<s:bean id="locales" name="com.xtgj.struts2.chapter08.locale.Locales" />
<form action="<s:url includeParams='get' encode='true'/>"
    name="langForm"
    style="background-color: powderblue; padding-top: 4px; padding-bottom: 4px;">
    Language:
    <s:select label="Language" list="#locales.locales" listKey="value"
        listValue="key"
        value="#SESSION_LOCALE == null ? locale : #SESSION_LOCALE"
        name="request_locale" id="langSelecter"
        onchange="langSelecter_onChanged()" theme="simple" />
</form>
```

上述代码的原理为，LangSelector.jsp 先实例化一个 Locales 对象，并把对象的 Map 类型的属性 locales 赋予下拉列表（select），因此下拉列表从而获得可用语言的列表。可以看到 LangSelector 有 <s:form> 标志和一段 JavaScript 脚本，它们的作用就是用户在下拉列表中选择后，提交包含“request_locale”变量的表单到 action。在打开页面时，在下拉列表选中的当前区域，我们需要到 session 取得当前区域(键为"WW_TRANS_I18N_LOCALE")的属性,而该属性在没有设置语言前为空,所以通过值栈中 locale 属性来取得当前区域(用户浏览器所设置的语言)。

你可以把 LangSelector.jsp 作为一个控件使用,方法是在 JSP 页面中把它包含进来,代码如下所示。

```
<s:include value="/LangSelector.jsp"/>
```

在示例 HelloWorld.jsp 中 <body> 后加入上述代码,并在 struts.xml 中新建 action,代码如下:

```
<action name="HelloWorld">
    <result>/HelloWorld.jsp</result>
</action>
```

或者,如果有多个 JSP 需要实现上述功能,可以使用下面的通用配置,而不是为每一个 JSP 页面都新建一个 action,如:

```
<action name="*">
    <result>/{1}.jsp</result>
</action>
```

发布运行程序,在浏览器的地址栏中输入 http://localhost:8080/Struts2_Chapter08/HelloWorld.action,出现如图 8-3 所示的页面。

在下拉列表中,选择"American English",出现和图 8-4 所示的页面。

图 8-3　中文输出

图 8-4　英文输出

可能大家会问为什么一定要通过 action 来访问页面呢?

你可以试一下不用 action 而直接用 JSP 的地址来访问页面,结果会是无论你在下拉列表中选择什么,语言都不会改变。这表示不能正常运行。其原因是如果直接使用 JSP 访问页面, Struts 2 在 web.xml 的配置的过滤器(Filter)就不会工作,所以拦截器链也不会工作。

8.2.4 包范围资源文件

在一个大型应用中，整个应用有大量的内容需要实现国际化，如果我们把国际化的内容都放置在全局资源属性文件中，显然会导致资源文件变得过于庞大、臃肿，不便于维护，这个时候我们可以针对不同模块，使用包范围来组织国际化文件。方法如下：

在 Java 的包下放置 package_language_country.properties 资源文件，package 为固定写法，处于该包及子包下的 action 都可以访问该资源。当查找指定 key 的消息时，系统会先从 package 资源文件查找，当找不到对应的 key 时，才会从常量 struts.custom.i18n.resources 指定的资源文件中寻找。

在上述案例的基础上将 globalMessages_en_US.properties 改为 package_en_US.properties 并放置于 com.xtgj.struts2.chapter08.locale 包下，将 globalMessages_zh_CN.properties 改为 package_zh_CN.properties 并放置于 com.xtgj.struts2.chapter08.locale 包下。在 HelloWorld.jsp 中添加代码：

```
<s:i18n name="com/xtgj/struts2/chapter08/locale/package">
    <h2>
        <s:text name="HelloWorld" />
    </h2>
</s:i18n>
```

在浏览器的地址栏中输入 http://localhost:8080/Struts2_Chapter08/HelloWorld.action，运行结果如图 8-5 所示。

图 8-5　包范围国际化

8.2.5 Action 范围资源文件

我们也可以为某个 action 单独指定资源文件，方法如下：

在 action 类所在的路径，放置 ActionClassName_language_country.properties 资源文件，ActionClassName 为 action 类的简单名称。

当查找指定 key 的消息时，系统会先从 ActionClassName_language_country.properties 资源文件查找，如果没有找到对应的 key，则会沿着当前包往上查找基本名为 package 的资源文件，一直找到最顶层包。如果还没有找到对应的 key，最后会从常量 struts.custom.i18n.resources 指定的资源文件中寻找。

在上述案例的基础添加资源文件 LocalesAction_en_US.properties 和 LocalesAction_zh_CN.properties，内容同上，并放置于和 LocalesAction 处在同一个位置的 com.xtgj.struts2.chapter08.locale 包下。

创建 action 类 LocalesAction.java，如示例代码 8-4 所示。

示例代码 8-4 LocalesAction.java 代码

```
package com.xtgj.struts2.chapter08.locale;

import com.opensymphony.xwork2.ActionContext;
import com.opensymphony.xwork2.ActionSupport;

public class LocalesAction extends ActionSupport {
    public String execute() {
        String text = this.getText("HelloWorld");
        ActionContext.getContext().put("text", text);
        return "success";
    }
}
```

在 HelloWorld.jsp 中添加代码：

```
<a href="HelloWorldRead.action"> 读取国际化信息 </a>
${text}
```

在 struts.xml 中添加 action 配置：

```
<action name="HelloWorldRead" class="com.xtgj.struts2.chapter08.locale.LocalesAction">
    <result>/HelloWorld.jsp</result>
</action>
```

在浏览器的地址栏中输入 http://localhost:8080/Struts2_Chapter08/HelloWorld.action，运行结果如图 8-6 所示。

图 8-6　Action 范围国际化

8.3　小结

✓ 在 Struts 2 中需要做国际化的有：jsp 页面的国际化、action 错误信息的国际化、转换错误信息的国际化、校验错误信息的国际化。

✓ 国际化资源文件的优先级：全局 < 包级别 < 类级别。

✓ 国际化信息的显示方法包括：

✧ <s:text name="HelloWorld"></s:text>

✧ getText("HelloWorld ")

✧ <message key=" HelloWorld "></message>

✧ <s:textfield name="username" key=" HelloWorld "></s:textfield>

✧ <s:i18n name=" HelloWorld "></s:i18n>

8.4　英语角

selector	选择器
internationalization	国际化
i18n	国际化简称

8.5　作业

回顾国际化的使用，使用全局国际化资源配置实现一个用户登录的业务。

8.6 思考题

Struts 2 中国际化是如何实现的？

8.7 学员回顾内容

国际化的使用。

上机部分

第 1 章　Struts 2 概述

本阶段目标

完成本章内容后，你将能够学到：

✧ Struts 2 的框架架构搭建。

✧ Struts 2 的控制器组件应用。

本阶段给出的步骤全面详细，请读者按照给出的上机步骤独立完成上机练习，以达到要求的学习目标。请认真完成下列步骤。

1.1　指导(1 小时 10 分钟)

搭建 Struts 2 环境

（1）从官方网下载 Struts 2 稳定版发布包（网址 http://struts.apache.org/download.cgi），这里我们下载 struts。

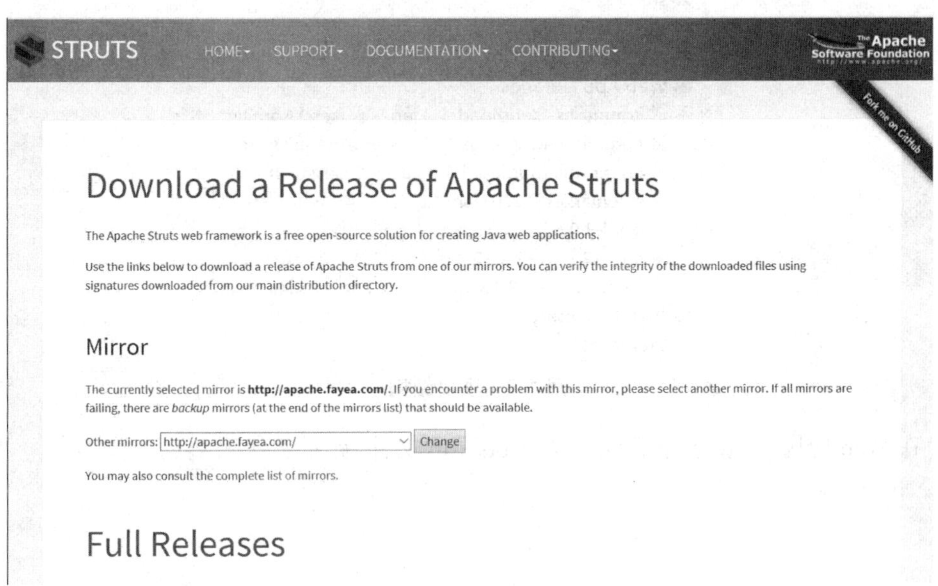

图 1-1　struts 下载界面

（2）在 MyEclipse 中新建一个 Java Web 项目，具体步骤参考图 1-2。

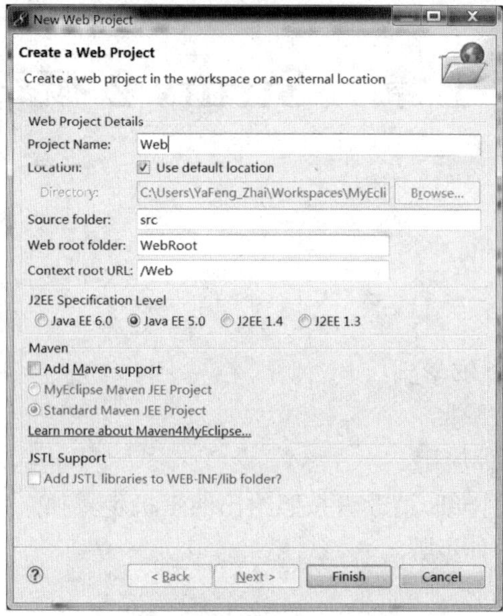

图 1-2 新建 Java web 项目窗口

（3）将下载的 struts-2.3.15.3 发布包解压，然后将 Struts 2 的必要 jar 包拷贝到 Web 应用的 lib 包中（其中 commons-fileupload-1.3.jar 和 commons-io-2.0.1.jar 主要用来操作文件的，并不是必需的）。图 1-3 为项目结构示意图。

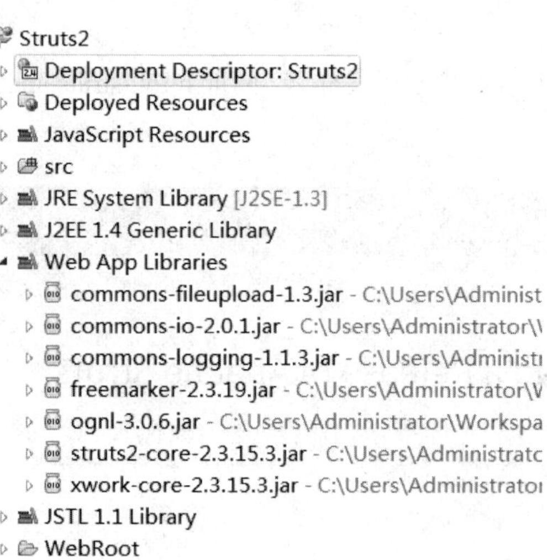

图 1-3 项目结构

（4）在 Web 应用的 src 包下新建一个 struts.xml 文件，如示例代码 1-1 所示。

示例代码 1-1　struts.xml 文件

```
<?xml version="1.0" encoding="UTF-8" ?>
<!DOCTYPE struts PUBLIC
    "-//Apache Software Foundation//DTD Struts Configuration 2.3//EN"
    "http://struts.apache.org/dtds/struts-2.3.dtd">
<struts>
    <!-- Add packages here -->
</struts>
```

（5）接下来修改 WEB-INF/web.xml 文件，配置 StrutsPrepareAndExecuteFilter 核心控制器。如示例代码 1-2 所示。

示例代码 1-2　配置核心控制器

```
<?xml version="1.0" encoding="UTF-8"?>
<web-app id="WebApp_9" version="2.4"
    xmlns="http://java.sun.com/xml/ns/j2ee"
    xmlns:xsi="http://www.w3.org/2001/XMLSchema-instance"
    xsi:schemaLocation="http://java.sun.com/xml/ns/j2ee http://java.sun.com/xml/ns/
j2ee/web-app_2_4.xsd">
    <display-name>Struts2</display-name>
    <filter>
        <filter-name>struts2</filter-name>
        <filter-class>
            org.apache.struts2.dispatcher.ng.filter.StrutsPrepareAndExecuteFilter
        </filter-class>
    </filter>
    <filter-mapping>
        <filter-name>struts2</filter-name>
        <url-pattern>/*</url-pattern>
    </filter-mapping>
    <welcome-file-list>
        <welcome-file>index.html</welcome-file>
    </welcome-file-list>
</web-app>
```

（6）至此，Struts 2 环境搭建完毕，接下来就可以正式开发一个 Struts 2 应用了。

（7）在 struts.xml 中添加如示例代码 1-3 所示的配置。

示例代码 1-3 在 struts.xml 中添加如下配置

```
<?xml version="1.0" encoding="UTF-8"?>
<!DOCTYPE struts PUBLIC
    "-//Apache Software Foundation//DTD Struts Configuration 2.3//EN"
    "http://struts.apache.org/dtds/struts-2.3.dtd">
<struts>
    <package name="test" namespace="/test" extends="struts-default">
        <action name="basicvalid"
            class="com.xtgj.struts2.chapter01.BasicvalidAction" method="execute">
            <result name="success">/msg.jsp</result>
            <result name="input">/index_new.jsp</result>
        </action>
    </package>
</struts>
```

（8）编写测试页面 index_new.jsp，如示例代码 1-4 所示。

示例代码 1-4 index_new.jsp

```
<%@ page language="java" import="java.util.*" pageEncoding="UTF-8"%>
<%@ taglib prefix="s" uri="/struts-tags"%>
<!DOCTYPE HTML PUBLIC "-//W3C//DTD HTML 4.01 Transitional//EN">
<html>
    <head>
        <title>My JSP 'index_new.jsp' starting page</title>
    </head>

    <body>
        <s:form method="post" action="test/basicvalid.action">
            <!-- 下面使用 Struts 2 标签定义三个表单域 -->
            <s:textfield label=" 名字 " name="name" />
            <s:textfield label=" 年纪 " name="age" />
            <s:textfield label=" 喜欢的颜色 " name="answer" />
            <!-- 定义一个提交按钮 -->
            <s:submit />
        </s:form>

    </body>
</html>
```

（9）编写测试成功导向的页面 msg.jsp，如示例代码 1-5 所示。

示例代码 1-5　msg.jsp

```
<%@ page language="java" import="java.util.*" pageEncoding="UTF-8"%>
<!DOCTYPE HTML PUBLIC "-//W3C//DTD HTML 4.01 Transitional//EN">
<html>
  <head>
    <title>My JSP 'msg.jsp' starting page</title>
  </head>
   <body>
    第一个 Struts2 项目测试成功！ <br>
    您填入的名字是：${name }<br>
  </body>
</html>
```

（10）编写验证 action，如示例代码 1-6 所示。

示例代码 1-6　验证 action

```
package com.xtgj.struts2.chapter01;
import com.opensymphony.xwork2.ActionSupport;
public class BasicvalidAction extends ActionSupport {
    private String name;
    public String getName() {
        return name;
    }
    public void setName(String name) {
        this.name = name;
    }
    public String execute() {
        System.out.println("execute method");
        return "success";
    }
    public void validate() {
        System.out.println("validate method");
    }
}
```

（11）编写验证文件，命名为 BasicvalidAction-validation.xml，并将其放置于 BasicvalidAction 类所在的包下，如示例代码 1-7 所示。

示例代码 1-7　BasicvalidAction-validation.xml

```
<?xml version="1.0" encoding="UTF-8"?>
<!DOCTYPE validators PUBLIC "-//OpenSymphony Group//XWork Validator 1.0.3//
EN" "http://www.opensymphony.com/xwork/xwork-validator-1.0.3.dtd">
<validators>
    <field name="name">
        <field-validator type="requiredstring">
            <param name="trim">true</param>
            <message> 用户名不能为空 !</message>
        </field-validator>
    </field>
</validators>
```

至此，一个简单的 Struts 2 项目就完成了。打开浏览器，在地址栏中输入“http://localhost:8080/ Struts2/ index_new.jsp”将会看到如图 1-4 的界面。

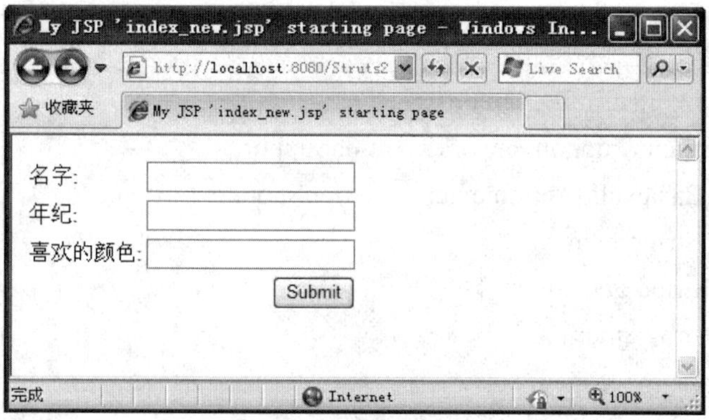

图 1-4　添加信息页面

若没有添加“名字”字段，点击“Submit”按钮时，会看到如图 1-5 的界面。

图 1-5　添加信息失败页面

若添加信息完整，例如填入名字为“tom”，点击“Submit”按钮时，会看到如图 1-6 的界面。

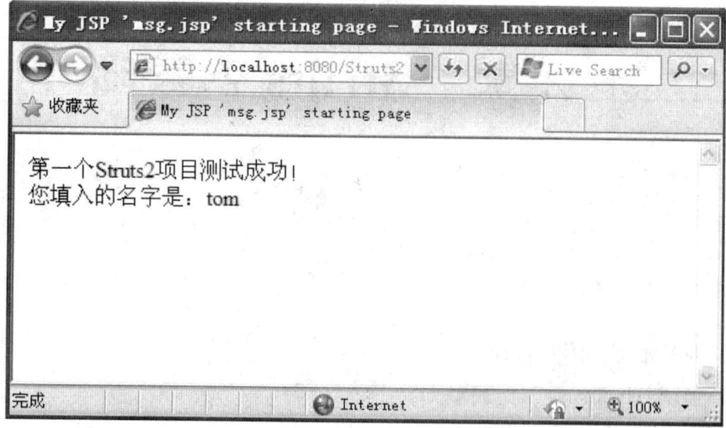

图 1-6　添加信息成功页面

1.2　作业

按照“1.1 指导”中的步骤搭建 Struts 2 环境，然后自己编写一个 HelloWorld 测试 Struts 2 环境搭建是否成功。

第 2 章　Struts 2 基础

本阶段目标

完成本章内容后，你将能够学到：

✧ 为 Web 应用增加 Struts 2 支持。

✧ Struts 2 框架的 MVC 组件。

✧ 使用 Struts 2 框架开发 Web 应用。

本阶段给出的步骤全面详细，请读者按照给出的上机步骤独立完成上机练习，以达到要求的学习目标，请认真完成下列步骤。

2.1　指导(1 小时 10 分钟)

深入 Struts 2 的配置文件

(1)struts.xml 文件配置灵活多样，现在给出一个简单示例演示 struts.xml 常用配置，如示例代码 2-1 所示。

示例代码 2-1　struts.xml

```
<?xml version="1.0" encoding="UTF-8" ?>
<!DOCTYPE struts PUBLIC
    "-//Apache Software Foundation//DTD Struts Configuration 2.3//EN"
    "http://struts.apache.org/dtds/struts-2.3.dtd">
<struts>
    <package name="example" extends="struts-default">
        <action name="Login_*" method="{1}" class="com.xtgj.action.Login">
            <result name="input">/login.jsp</result>
            <result name="error" type="redirectAction">error</result>
            <result>/result.jsp</result>
        </action>
        <action name="*" class="com.xtgj.action.TestAction">
```

```
            <result>/{1}.jsp</result>
        </action>
    </package>
</struts>
```

（2）在 struts.xml 配置中的 <action> 有些常用的默认值，例如：

① <action> 中如果没有指定 class 属性值，那么其默认值为 ActionSupport；

② <action> 中如果没有指定 method 属性值，那么其默认值为 execute；

③ <action> 中的 <result> 如果没有指定 name 属性值，那么其默认值为 success。

示例代码 2-2　struts.xml 代码片段

```
<action name="error" class="com.xtgj.action.TestAction">
            <result>/error.jsp</result>
</action>
```

这里请求会自动转发到 TestAction 类中的 execute() 方法中，经过处理之后，如果返回“success”，那么将导向 error.jsp 页面。

（3）<result> 有多种转发类型：

① dispatcher 为默认值，表示请求转发；

② redirect 表示请求重定向；

③ redirectAction 表示将请求重定向到另一个 Action 中；

④ plainText 表示显示原始文件内容。

示例代码 2-3　result 配置示例

```
<result name="error" type="redirectAction">error</result>
```

这里表示将请求重定向到同包下的另一个名为“error”的 action 中。

（4）Struts 2 允许使用通配符定义 action，如示例代码 2-4 所示。

示例代码 2-4　使用通配符定义 action

```
……
<action name="Login_*" method="{1}" class="com.xtgj.action.Login">
    ……
</action>
<action name="*" class="com.xtgj.action.TestAction">
    <result>/{1}.jsp</result>
</action>
……
```

这里名为“Login_*”的 action 表示可以接受所有以“Login”开头的请求，后面的“method={1}”中 {1} 是一个占位符，由请求“Login_*”中的“*”号部分替代。假如，有“Login_input.action”的请求发出，那么这个请求将会被转到 Login action 中的 input() 方法去处理。类似的，名为“*”的 action 代表可以接受任意未经匹配的请求。

2.2 练习 1(30 分钟)

利用“2.1 指导”，写一个登录的例子，具体要求如下：

第一步，编写首页 welcome.jsp，界面如图 2-1 所示。

图 2-1 首页界面

其代码如示例代码 2-5 所示。

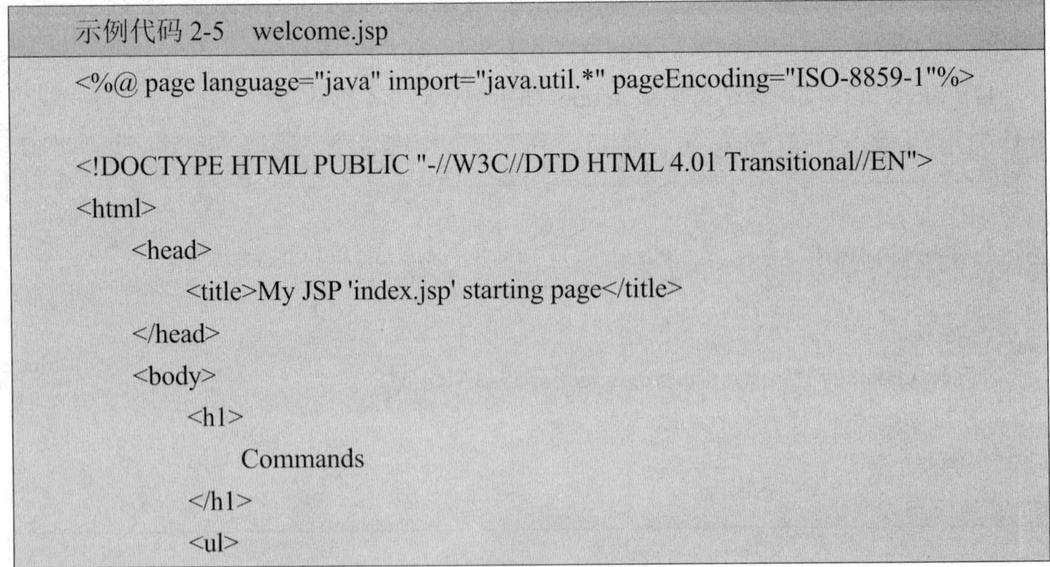

示例代码 2-5 welcome.jsp

```
<%@ page language="java" import="java.util.*" pageEncoding="ISO-8859-1"%>

<!DOCTYPE HTML PUBLIC "-//W3C//DTD HTML 4.01 Transitional//EN">
<html>
    <head>
        <title>My JSP 'index.jsp' starting page</title>
    </head>
    <body>
        <h1>
            Commands
        </h1>
        <ul>
```

```
                <li>
                    <a href="Login.jsp">Sign on</a>
                </li>
                <li>
                    <a href="Register.jsp">Register</a>
                </li>
            </ul>
        </body>
</html>
```

第二步，编写登录页面。当点击图 2-1 上的“Sign on”链接后，页面导向登录界面 login.jsp，如图 2-2 所示。其代码如示例代码 2-6 所示。

图 2-2　登录界面

示例代码 2-6　login.jsp

```
<%@ page language="java" import="java.util.*" pageEncoding="UTF-8"%>
<!DOCTYPE HTML PUBLIC "-//W3C//DTD HTML 4.01 Transitional//EN">
<html>
    <head>
        <title>My JSP 'Login.jsp' starting page</title>
    </head>
    <body>
        <table border="1" cellspacing="0">
            <tr>
                <td bgcolor="pink">
                    <table border="0" cellspacing="0" cellpadding="4">
```

```
                    <tr>
                        <td>
                            <font    color="#CCCCCC">     <font
color="#FFFFFF"> 用户登录 </font>
                            </font>
                        </td>
                    </tr>
                </table>
            </td>
        </tr>
        <tr>
            <td bgcolor="#EAEAEA" colspan="2">
                <form name="userForm" action="user_login.action" method="post">
                    <p>
                        <label for="textfield1">
                            用户名:
                        </label>
                        <input type="text" name="username" id="username">
                    </p>
                    <p>
                        <label for="textfield2">
                            密      码:
                        </label>
                        <input       type="password"       name="password"
id="password">
                    </p>
                    <p align="center">
                        <input type="submit" name="Submit" value=" 登录 ">
                    </p>
                </form>
            </td>
        </tr>
    </table>
  </body>
</html>
```

输入用户名（scott）和密码（tiger）后，请求仍然发送到 LoginAction，只不过是由其中的 execute() 方法进行处理，从而进入结果页面 result.jsp，并回显用户名，如图 2-3 所示。

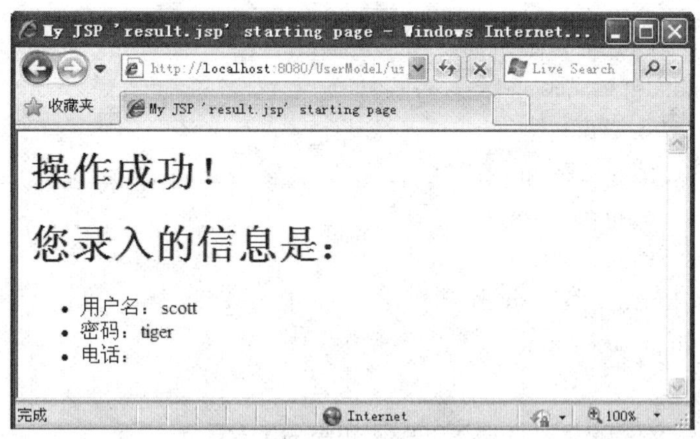

图 2-3　登录成功界面

登录成功页面 result.jsp 的代码如示例代码 2-7 所示。

示例代码 2-7　result.jsp

```
<%@ page language="java" import="java.util.*" pageEncoding="UTF-8"%>
<!DOCTYPE HTML PUBLIC "-//W3C//DTD HTML 4.01 Transitional//EN">
<html>
  <head>
    <title>My JSP 'result.jsp' starting page</title>
  </head>
  <body>
    <h1> 操作成功！</h1>
    <h1> 您录入的信息是:</h1>
    <ul>
                <li>
                    用户名:${username }
                </li>
                <li>
                    密码:${password }
                </li>
                <li>
                    电话:${phone }
                </li>
        </ul>
  </body>
</html>
```

第三步，编写处理登录请求的 action。UserAction 代码如示例代码 2-8 所示。

示例代码 2-8　编写处理登录请求的 Action

```
package com.xtgj.user.action;
public class UserAction {
	private String username;
	private String password;
	private String phone;
	public String getUsername() {
		return username;
	}
	public void setUsername(String username) {
		this.username = username;
	}
	public String getPassword() {
		return password;
	}
	public void setPassword(String password) {
		this.password = password;
	}
	public String getPhone() {
		return phone;
	}
	public void setPhone(String phone) {
		this.phone = phone;
	}
	public String login() {
		if (username.equals("scott") && password.equals("tiger")) {
			return "success";
		} else {
			return "error";
		}
	}
	public String register() {
		if (username != null && !username.equals("")
				&& !username.equals("scott")) {
			return "success";
		} else {
			return "error";
		}
```

```
        }
    }
```

第四步，编写报错页面 error.jsp。当输入其他用户名和密码后，请求直接跳转到 TestAction 中，不做任何处理直接进入错误页面 error.jsp，回显信息为“对不起，该网页正在升级中……”，如图 2-4 所示。

图 2-4　登录失败界面

error.jsp 代码如示例代码 2-9 所示。

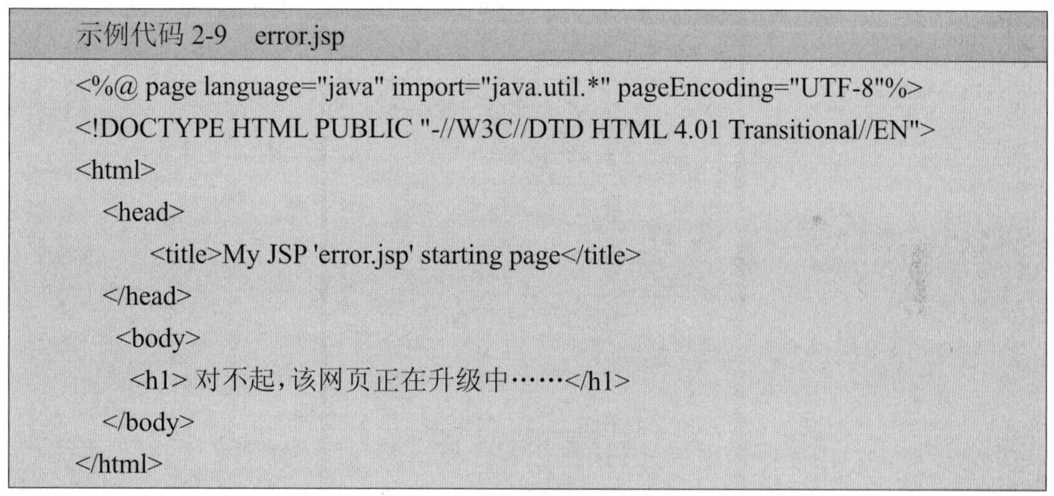

示例代码 2-9　error.jsp

```
<%@ page language="java" import="java.util.*" pageEncoding="UTF-8"%>
<!DOCTYPE HTML PUBLIC "-//W3C//DTD HTML 4.01 Transitional//EN">
<html>
  <head>
      <title>My JSP 'error.jsp' starting page</title>
  </head>
   <body>
     <h1> 对不起，该网页正在升级中……</h1>
  </body>
</html>
```

第五步，配置 struts.xml。struts.xml 代码如示例代码 2-10 所示。

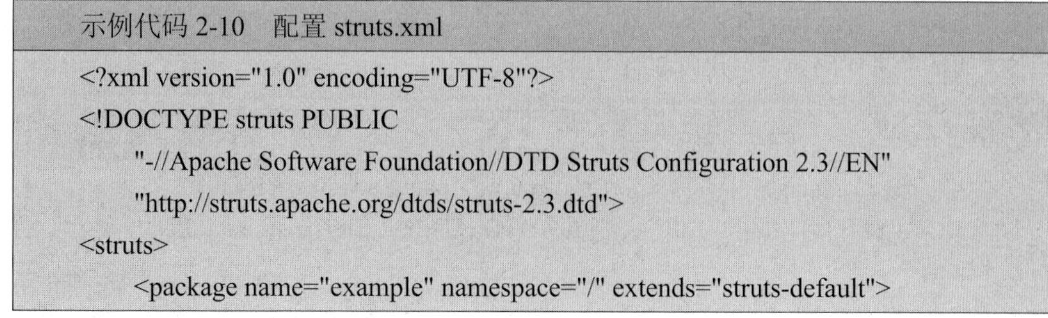

示例代码 2-10　配置 struts.xml

```
<?xml version="1.0" encoding="UTF-8"?>
<!DOCTYPE struts PUBLIC
    "-//Apache Software Foundation//DTD Struts Configuration 2.3//EN"
    "http://struts.apache.org/dtds/struts-2.3.dtd">
<struts>
    <package name="example" namespace="/" extends="struts-default">
```

```
            <action name="user_*" method="{1}" class="com.xtgj.user.action.UserAction">
                <result>/result.jsp</result>
                <result name="error">/error.jsp</result>
            </action>
        </package>
</struts>
```

2.3 练习 2(30 分钟)

继“2.2 练习”后，实现首页中的“Register”链接功能，不过该功能要求，只要点击“Register”链接，将导向 Register.jsp 页面，其回显信息仍然是“对不起，该网页正在升级中……”，这里要求不添加任何额外的 action 或配置代码来实现，注册页面如图 2-5 所示。

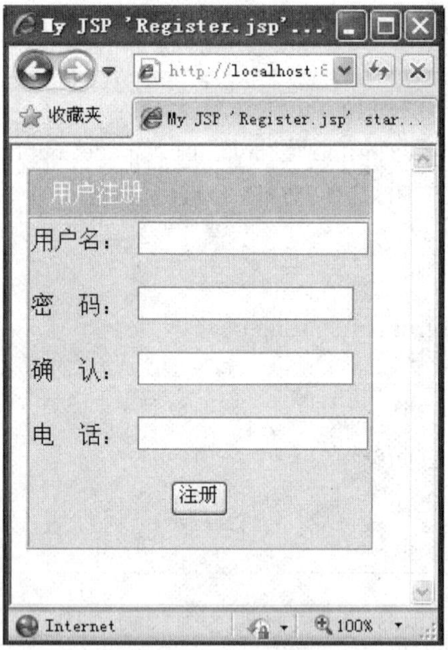

图 2-5 注册界面

注册页面参考代码如示例代码 2-11 所示。

示例代码 2-11 Register.jsp

```
<%@ page language="java" import="java.util.*" pageEncoding="UTF-8"%>
<!DOCTYPE HTML PUBLIC "-//W3C//DTD HTML 4.01 Transitional//EN">
<html>
  <head>
```

```
        <title>My JSP 'Register.jsp' starting page</title>
    </head>
    <body>
            <table border="1" cellspacing="0">
                <tr>
                    <td bgcolor="pink">
                        <table border="0" cellspacing="0" cellpadding="4">
                            <tr>
                                <td>
                                    <font  color="#CCCCCC">   <font
color="#FFFFFF"> 用户注册 </font>
                                    </font>
                                </td>
                            </tr>
                        </table>
                    </td>
                </tr>
                <tr>
                    <td bgcolor="#EAEAEA" colspan="2">
                        <form  name="userForm"  action="user_register"  method="
post">
                            <p>
                                <label for="textfield1">
                                    用户名:
                                </label>
                                <input  type="text"  name="username"  id="user-
name">
                            </p>
                            <p>
                                <label for="textfield2">
                                    密      码:
                                </label>
                                <input  type="password"  name="password"  id="-
password">
                            </p>
                            <p>
                                <label for="textfield2">
                                    确      认:
```

```
                                        </label>
                                        <input    type="password"    name="password1"
id="password1">
                                    </p>

                                    <p>
                                        <label for="textfield2">
                                            电      话:
                                        </label>
                                        <input type="text" name="phone" id="phone">
                                    </p>
                                    <p align="center">
                                        <input type="submit" name="Submit" value=" 注
册 ">

                                    </p>
                                </form>
                            </td>
                        </tr>
                </table>
            </body>
        </html>
```

在如图 2-5 所示的页面中录入注册信息，并点击注册按钮，注册成功页面如图 2-6 所示。

图 2-6 注册成功界面

2.4　作业

根据用户登录和注册的业务需求分析，创建用户表 Users，如表 2-1 所示，实现用户登录和注册的具体业务功能。

表 2-1　Users 表

字段名	类型	是否空	是否主键
userid	varchar(20)	否	是
username	varchar(20)	否	否
password	varchar(20)	否	否
phone	varchar(20)	是	否

第 3 章　深入了解 Struts 2

本阶段目标

完成本章内容后，你将能够学到：

✧ Struts 2 中的多模块划分。

✧ Struts 2 中的全局 result 配置。

✧ 如何为 action 的属性注入值。

✧ 如何指定 Struts 2 处理的请求后缀。

✧ 如何使用通配符定义 Action。

✧ Struts 2 中的文件上传。

本阶段给出的步骤全面详细，请学员按照给出的上机步骤独立完成上机练习，以达到要求的学习目标。请认真完成下列步骤。

3.1　指导(1 小时 10 分钟)

3.1.1　Struts 2 中的多模块划分

在大部分应用里，随着应用规模的增加，系统中 action 的数量也会大量增加，导致 struts.xml 配置文件变得非常臃肿。为了避免 struts.xml 文件过于庞大、臃肿，提高 struts.xml 文件的可读性，我们可以将一个 struts.xml 配置文件分解成多个配置文件，然后在 struts.xml 文件中包含其他配置文件。下例的 struts.xml 通过 <include> 元素指定多个配置文件。

struts.xml 文件配置灵活多样，通过 <include> 元素指定多个配置文件，如示例代码 3-1 所示。

示例代码 3-1　include 配置示例

```
<?xml version="1.0" encoding="UTF-8"?>
<!DOCTYPE struts PUBLIC
    "-//Apache Software Foundation//DTD Struts Configuration 2.3//EN"
    "http://struts.apache.org/dtds/struts-2.3.dtd">
<struts>
    <constant name="struts.action.extension" value="action,do" />
    <include file="dept.xml" />
    <include file="product.xml" />
</struts>
```

通过这种方式，我们就可以将 Struts 2 的 action 按模块添加在多个配置文件中。上述代码中的“<constant name="struts.action.extension" value="action,do" />”说明 Struts 2 处理的请求后缀可以是“action”或“do”。上述代码中的 dept.xml 和 product.xml 是符合“struts-2.3.dtd”的 XML 文档。

dept.xml 的代码如示例代码 3-2 所示。

示例代码 3-2　dept.xml

```
<?xml version="1.0" encoding="UTF-8"?>
<!DOCTYPE struts PUBLIC
    "-//Apache Software Foundation//DTD Struts Configuration 2.3//EN"
    "http://struts.apache.org/dtds/struts-2.3.dtd">
<struts>
    <package name="dept" namespace="/dept" extends="struts-default">
        <global-results>
            <result name="msg">msg.jsp</result>
        </global-results>
        <action name="dept_*" class="com.xtgj.struts2.dept.DeptAction" meth-
od="{1}">
            <result name="list">deptList.jsp</result>
            <result name="updateUI">deptUpdateUI.jsp</result>
            <result name="addUI">deptAddUI.jsp</result>
        </action>
    </package>
</struts>
```

product.xml 的代码如示例代码 3-3 所示。

示例代码 3-3 product.xml

```
<?xml version="1.0" encoding="UTF-8"?>
<!DOCTYPE struts PUBLIC
    "-//Apache Software Foundation//DTD Struts Configuration 2.3//EN"
    "http://struts.apache.org/dtds/struts-2.3.dtd">
<struts>
    <package name="product" namespace="/product"
        extends="struts-default">
        <global-results>
            <result name="msg">msg.jsp</result>
        </global-results>
        <action name="pro_*" class="com.xtgj.struts2.product.ProductAction"
            method="{1}">

        </action>
    </package>
</struts>
```

如果 action 中存在多个方法时，我们可以使用通配符方式定义 action，例如：

（1）“dept_*”表示所有前缀是“dept_”的 action 请求都可以通配到指定的 DeptAction 中，具体调用 action 中的哪一个方法，可以通过指定 method 的值决定，例如 method 的值为“{1}”，则表示调用“dept_*”这个 action 中和第一个“*”号通配的名称相同名称的方法。

（2）“pro_*”表示所有前缀是“pro_”的 action 请求都可以通配到指定的 ProductAction 中，具体调用 action 中的哪一个方法，可以通过指定 method 的值决定，例如 method 的值为“{1}”，则表示调用“pro _*”这个 action 中和第一个“*”号通配的名称相同名称的方法。

这样，定义的 action 中可以存在多个方法，操作灵活，减少了不必要的代码，使得项目结构简单、整洁。我们在访问某个 action 时，可以使用类似下述方式：

```
http://localhost:8080/ProductManagement/dept/dept_add
```

这样，应用程序会访问 dept 模块中的“dept_add”请求，并调用 DeptAction 中的 add() 方法。

在 struts 2 框架中使用包来管理 action。在实际应用中，我们应该把一组业务功能相关的 action 放在同一个包下。包的 namespace 属性用于定义该包的命名空间，命名空间作为访问该包下 action 的路径的一部分，事实上，这也是多模块划分的一个重要体现。通常我们可以将一组视图放在对应命名空间名称命名的文件夹中，即一个模块。仍然以“http://localhost:8080/ProductManagement/dept/dept_add”这个访问路径为例，我们可以认为所有的“dept_*”请求被划分到了“dept”模块下。所以，在创建视图组件时，应该注意模块的划分。本项目中的视图组件被分为两个模块，分别是部门管理模块和商品管理模块，因此需要在该项目 WebRoot 目录下创建一个“dept”文件夹和一个“product”文件夹。

3.1.2　Struts 2 中的复合类型参数

（1）Struts 2 采用复合类型接收请求参数，是指 Struts 2 首先通过反射技术调用某个类的默认构造器创建该类的对象，然后再通过反射技术调用该对象中与请求参数同名的属性的 set 方法来获取请求参数值。

本项目中的 Dept 域类代码如示例代码 3-4 所示。

示例代码 3-4　Dept 域类

```
package com.xtgj.struts2.domain;
import java.util.Date;
public class Dept {
    private Integer deptno;// 部门编号
    private String dname; // 部门名称
    private Date creationTime; // 创建时间
    public Dept(Integer deptno, String dname, Date creationTime) {
        this.deptno = deptno;
        this.dname = dname;
        this.creationTime = creationTime;
    }
    public Dept() {
    }
    public Integer getDeptno() {
        return deptno;
    }
    public void setDeptno(Integer deptno) {
        this.deptno = deptno;
    }
    public String getDname() {
        return dname;
    }
    public void setDname(String dname) {
        this.dname = dname;
    }
    public Date getCreationTime() {
        return creationTime;
    }
    public void setCreationTime(Date creationTime) {
```

```
            this.creationTime = creationTime;
        }
}
```

（2）DeptAction 代码如示例代码 3-5 所示，注意 action 中的 dept 属性。

示例代码 3-5　DeptAction

```
package com.xtgj.struts2.dept;
import java.util.ArrayList;
import java.util.Date;
import java.util.Iterator;
import java.util.List;
import com.opensymphony.xwork2.ActionContext;
import com.xtgj.struts2.domain.Dept;
public class DeptAction {
        private Dept dept;// 部门对象模型
        private static List<Dept> list = new ArrayList<Dept>(); // 模拟部门初始信息
        // 一个静态块初始化 list 集合中的元素
        static {
                list.add(new Dept(1, "SALE", new Date()));
                list.add(new Dept(2, "CLERK", new Date()));
                list.add(new Dept(3, "PERSON", new Date()));
        }
        public String list() {
                 return "list";
        }
        public String addUI() {
                return "addUI";
        }
        public String add() {
                list.add(new Dept(dept.getDeptno(), dept.getDname(), dept
                                .getCreationTime()));
                return "list";
        }
        public String del() {
                Integer dno = dept.getDeptno();
                Iterator<Dept> it = list.iterator();
                Dept tmp = null;
```

```
        while (it.hasNext()) {
            tmp = it.next();
            if (tmp.getDeptno().equals(dno)) {
                list.remove(tmp);
                break;
            }
        }
        return "list";
    }
    public String updateUI() {
        Integer dno = dept.getDeptno();
        Iterator<Dept> it = list.iterator();
        Dept tmp = null;
        while (it.hasNext()) {
            tmp = it.next();
            if (tmp.getDeptno().equals(dno)) {
                dept = tmp;
                break;
            }
        }
        ActionContext ctx = ActionContext.getContext();
        ctx.put("dept", dept);
        return "updateUI";
    }
    public String update() {
        Integer dno = dept.getDeptno();
        Iterator<Dept> it = list.iterator();
        Dept tmp = null;
        while (it.hasNext()) {
            tmp = it.next();
            if (tmp.getDeptno().equals(dno)) {
                list.remove(tmp);
                break;
            }
        }
        list.add(new Dept(dept.getDeptno(), dept.getDname(), dept
                .getCreationTime()));
        return "list";
```

```
        }
        public List<Dept> getList() {
                return list;
        }
        public Dept getDept() {
                return dept;
        }
        public void setDept(Dept dept) {
                this.dept = dept;
        }
}
```

在这个类中封装了最基本的 CRUD 操作，但是未实现 DAO 操作，请读者根据基本业务需求完成真正的 CRUD 操作，并进行实验。

（3）部门添加页面代码如示例代码 3-6 所示。

示例代码 3-6　部门添加页面

```
<%@ page language="java" import="java.util.*" pageEncoding="UTF-8"%>
<!DOCTYPE HTML PUBLIC "-//W3C//DTD HTML 4.01 Transitional//EN">
<html>
        <head>
                <title>My JSP 'deptAddUI.jsp' starting page</title>
        </head>
        <body>
                <center>
                        <h2>
                                添加部门
                        </h2>
                        <br />
                        <form action="dept_add" method="post">
                                部门编号：
                                <input name="dept.deptno" value="" type="text" />
                                <br />
                                部门名称：
                                <input name="dept.dname" value="" type="text" />
                                <br />
                                创建时间：
                                <input name="dept.creationTime" value="" type="text" />
```

```
                    <br />
                    <input value=" 添加 " type="submit" /><input value=" 重置 " type="reset" />
                    <br />
                </form>
            </center>
        </body>
    </html>
```

其效果图如图 3-1 所示。

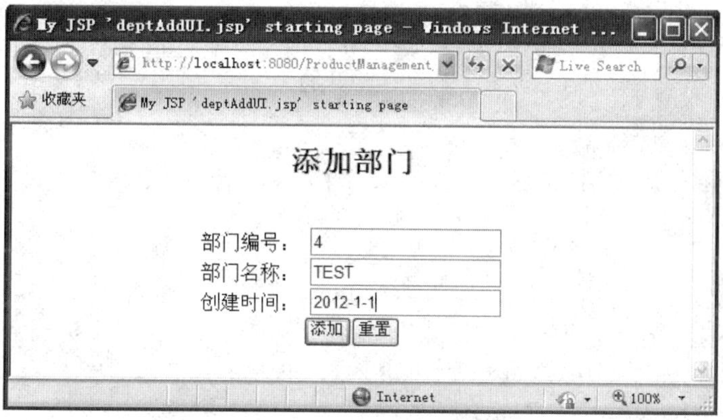

图 3-1　部门添加页面

（4）部门信息列表页面代码如示例代码 3-7 所示。

示例代码 3-7　部门信息列表页面

```
<%@ page language="java" import="java.util.*" pageEncoding="UTF-8"%>
<%@taglib prefix="c" uri="http://java.sun.com/jsp/jstl/core"%>
<!DOCTYPE HTML PUBLIC "-//W3C//DTD HTML 4.01 Transitional//EN">
<html>
    <head>
        <title>My JSP 'deptList.jsp' starting page</title>
  </head>

    <body>
        <center>
            <h2> 部门列表 </h2>
            <a href="dept_addUI"> 增加部门 </a>

        </center>
```

```
        <br>
        <table align="center" border="1">
        <tr>
                    <td>
                        部门号
                    </td>
                    <td>
                        部门名
                    </td>
                    <td>
                        部门创建时间
                    </td>
                    <td colspan="2">
                        操作
                    </td>
                </tr>

            <c:forEach items="${list}" var="dept">
                <tr>
                    <td>
                        <c:out value="${dept.deptno}" />

                    </td>
                    <td>
                        <c:out value="${dept.dname}" />
                    </td>
                    <td>
                        <c:out value="${dept.creationTime}" />
                    </td>
                    <td>
                        <a  href="dept_del.action?dept.deptno=${dept.deptno}"> 删
除 </a>
                    </td>
                    <td>
                        <a  href="dept_updateUI?dept.deptno=${dept.deptno}"> 修
改 </a>
                    </td>
                </tr>
```

```
            </c:forEach>
        </table>
    </body>
</html>
```

其效果图如图 3-2 所示。

图 3-2　部门列表页面

在该页面上点击删除链接，将该记录删除，若删除成功，仍然导向部门列表页面；在该页面上点击修改链接，进入修改信息页面，修改后提交修改请求，若修改成功，仍然导向部门列表页面。

（5）部门修改页面代码如示例代码 3-8 所示。

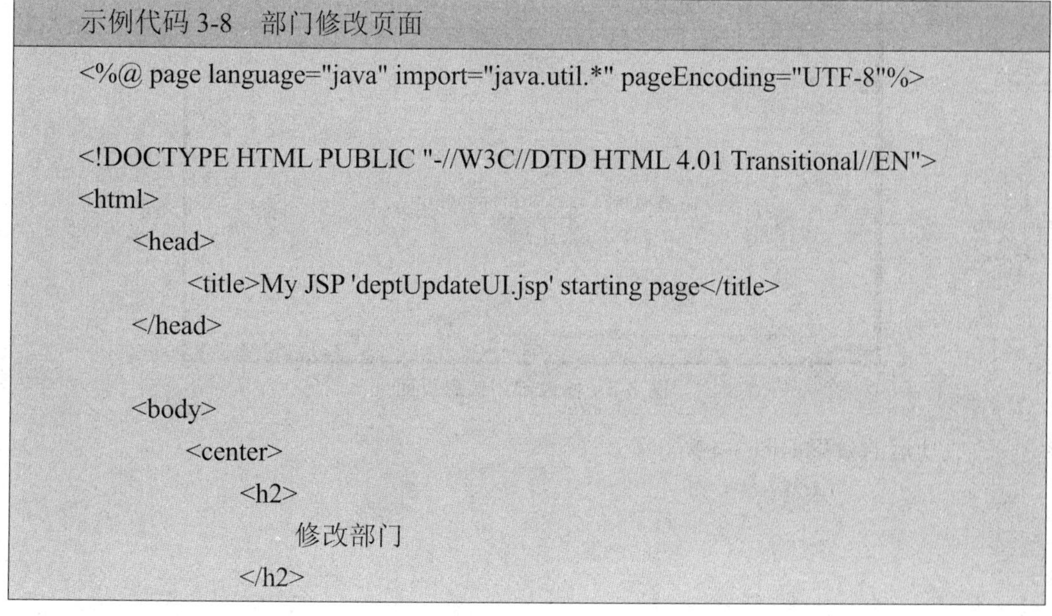

示例代码 3-8　部门修改页面

```
<%@ page language="java" import="java.util.*" pageEncoding="UTF-8"%>

<!DOCTYPE HTML PUBLIC "-//W3C//DTD HTML 4.01 Transitional//EN">
<html>
    <head>
        <title>My JSP 'deptUpdateUI.jsp' starting page</title>
    </head>

    <body>
        <center>
            <h2>
                修改部门
            </h2>
```

```
                <br />
                <form action="dept_update" method="post">
                    部门编号：
                    <input name="dept.deptno" value="${dept.deptno}" type="text" />
                    <br />
                    部门名称：
                    <input name="dept.dname" value="${dept.dname}" type="text" />
                    <br />
                    创建时间：
                    <input name="dept.creationTime" value="${dept.creationTime}"
                        type="text" />
                    <br />
                    <input value=" 修改 " type="submit" /><input value=" 重置 "
type="reset" />
                    <br />
                </form>
            </center>
        </body>
    </html>
```

其效果如图 3-3 所示。

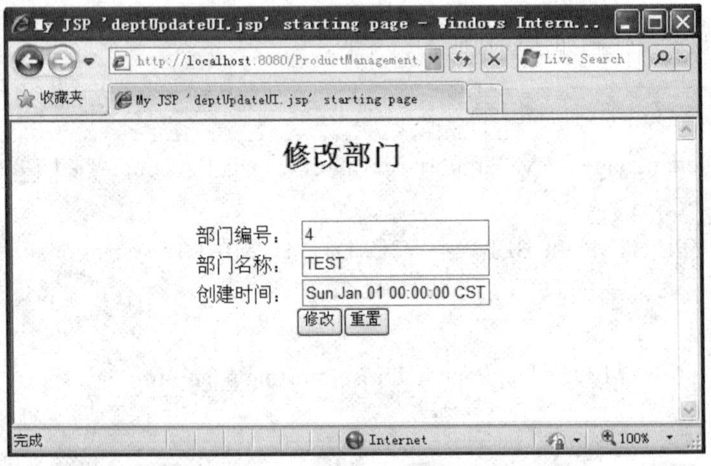

图 3-3 修改部门信息页面

修改成功后的效果如图 3-4 所示。

图 3-4　修改成功示意图

3.2　练习（30 分钟）

利用“3.1 指导”，继续完成 product 模块，重点实现文件上传。这里总结文件上传的三个步骤：

- 在 WEB-INF/lib 下加入 commons-fileupload-1.3.jar、commons-io-2.0.1.jar。这两个文件可以从 http://commons.apache.org/ 下载；
- 把 form 表单的 enctype 属性值设置为“multipart/form-data”，例如：

```
<form enctype="multipart/form-data" action="xxx.action" method="post">
   <input type="file" name="uploadImage"/>
</form>
```

- 在对应表单的域类中添加文件对象属性、文件类型属性和文件名称属性，注意文件对象属性命名应同表单中文件字段名相同，文件类型和文件名称属性的命名要求必须以文件对象属性名为前缀。例如：

```
public class FileModel{
      private File uploadImage;// 文件对象
      private String uploadImageContentType;// 文件类型
      private String uploadImageFileName;// 文件名称
      // 这里略省了属性的 get/set 方法
}
```

第一步，编写商品添加页面“proAddUI.jsp”，界面如图 3-5 所示。

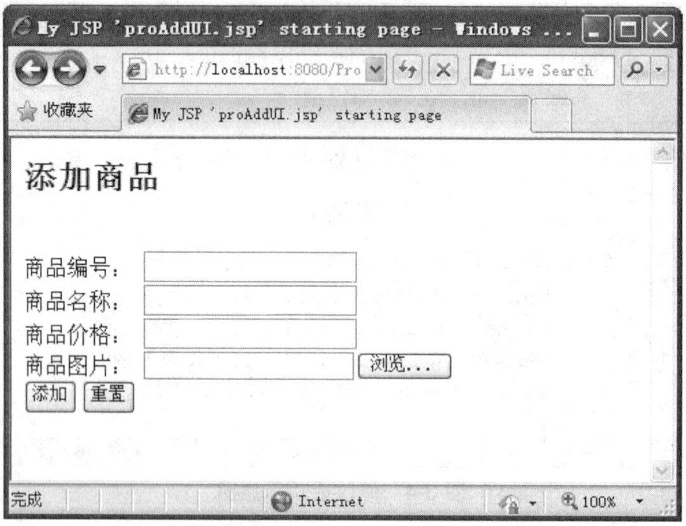

图 3-5 商品添加页面

其代码如示例代码 3-9 所示。

示例代码 3-9 proAddUI.jsp

```
<%@ page language="java" import="java.util.*" pageEncoding="UTF-8"%>
<!DOCTYPE HTML PUBLIC "-//W3C//DTD HTML 4.01 Transitional//EN">
<html>
    <head>
        <title>My JSP 'proAddUI.jsp' starting page</title>
    </head>
    <body>
        <h2>
            添加商品
        </h2>
        <br />
        <form action="pro_add" method="post" enctype="multipart/form-data">
            商品编号:
            <input name="product.pid" value="" type="text" />
            <br />
            商品名称:
            <input name="product.name" value="" type="text" />
            <br />
            商品价格:
            <input name="product.price" value="" type="text" />
            <br />
```

```
            商品图片：
            <input name="product.picPath" type="file" />
            <br />
            <input value=" 添加 " type="submit" />
            <input value=" 重置 " type="reset" />
            <br />
        </form>
    </body>
</html>
```

第二步，编写 Product 域类。Product 类代码如示例代码 3-10 所示。

示例代码 3-10　Product 类

```
package com.xtgj.struts2.domain;
import java.io.File;
public class Product {
    private Integer pid; // 商品编号
    private String name; // 商品名称
    private File picPath; // 图片路径
    private double price; // 商品单价
    private String picPathContentType; // 图片类型
    private String picPathFileName; // 图片名称
    public Product() {
    }
    public Product(Integer pid, String name, double price) {
        this.pid = pid;
        this.name = name;
        this.price = price;
    }
    public Integer getPid() {
        return pid;
    }
    public void setPid(Integer pid) {
        this.pid = pid;
    }
    public String getName() {
        return name;
    }
```

```
        public void setName(String name) {
            this.name = name;
        }
        public double getPrice() {
            return price;
        }
        public void setPrice(double price) {
            this.price = price;
        }
            public File getPicPath() {
            return picPath;
        }

            public void setPicPath(File picPath) {
            this.picPath = picPath;
        }
        public String getPicPathContentType() {
            return picPathContentType;
        }
        public void setPicPathContentType(String picPathContentType) {
            this.picPathContentType = picPathContentType;
        }
        public String getPicPathFileName() {
            return picPathFileName;
        }
            public void setPicPathFileName(String picPathFileName) {
            this.picPathFileName = picPathFileName;
        }
    }
```

第三步，编写处理商品添加请求的 action。ProductAction 代码如示例代码 3-11 所示。

示例代码 3-11 编写处理商品添加请求的 action

```
package com.xtgj.struts2.product;
import java.io.File;
import org.apache.commons.io.FileUtils;
import org.apache.struts2.ServletActionContext;
import com.opensymphony.xwork2.ActionContext;
```

```
import com.xtgj.struts2.domain.Product;
public class ProductAction {
    private Product product;
    public Product getProduct() {
        return product;
    }
    public void setProduct(Product product) {
        this.product = product;
    }
    public String add() throws Exception {
        System.out.println(product.getName() + "," + product.getPrice() + ","
                + product.getPicPathContentType() + ","
                + product.getPicPathFileName());
        String realpath = ServletActionContext.getServletContext().getRealPath(
                "/images");
          File file = new File(realpath);
          if (!file.exists())
               file.mkdirs();
        FileUtils.copyFile(product.getPicPath(), new File(file, product
                .getPicPathFileName()));
        ActionContext.getContext()
               .put("filename", product.getPicPathFileName());
        return "msg";
    }
}
```

第四步，编写商品添加成功后的回显页面 msg.jsp，如图 3-6 所示。

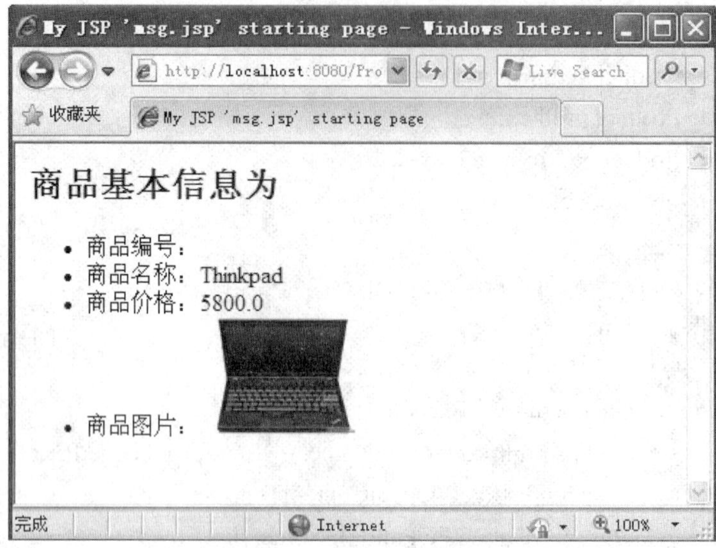

图 3-6 商品添加成功页面

msg.jsp 代码如示例代码 3-12 所示。

示例代码 3-12 msg.jsp

```
<%@ page language="java" import="java.util.*" pageEncoding="UTF-8"%>
<%
    String path = request.getContextPath();
    String basePath = request.getScheme() + "://"
            + request.getServerName() + ":" + request.getServerPort()
            + path + "/";
%>
<!DOCTYPE HTML PUBLIC "-//W3C//DTD HTML 4.01 Transitional//EN">
<html>
    <head>
        <base href="<%=basePath%>">
        <title>My JSP 'msg.jsp' starting page</title>
        <meta http-equiv="pragma" content="no-cache">
        <meta http-equiv="cache-control" content="no-cache">
        <meta http-equiv="expires" content="0">
    </head>
    <body>
        <h2>
            商品基本信息为
        </h2>
        <ul>
```

```
                <li>
                    商品编号:${product.pid }
                </li>
                <li>
                    商品名称:${product.name }
                </li>
                <li>
                    商品价格:${product.price }
                </li>
                <li>
                    商品图片:
    <img src="images/${filename }" width="100" height="80" alt=" 图片未能显示 " />
                </li>
            </ul>
        </body>
    </html>
```

第五步，配置 struts.xml。通过 <include> 元素指定 product 模块，具体方式请参考部门管理模块。

至此，我们就完成了商品管理模块中的商品信息添加功能，请读者阅读并实战。

3.3　作业

根据部门管理和商品管理的业务需求分析，创建部门表 dept，如表 3-1 所示，商品表 product，如表 3-2 所示，实现部门 CRUD 和商品 CRUD 操作。

表 3-1　dept 表

字段名	类型	是否空	是否主键
deptno	int	否	是
dname	varchar(20)	否	否
creationTime	datetime	否	否

表 3-2　product 表

字段名	类型	是否空	是否主键
pid	int	否	是
name	varchar(20)	否	否
picPath	varchar(50)	否	否
price	money	否	否

第 4 章　Struts 2 转换器

本阶段目标

完成本章内容后，你将能够学到：

✧ 转换器的组成结构。

✧ 转换器的使用。

本阶段给出的步骤全面详细，请学员按照给出的上机步骤独立完成上机练习，以达到要求的学习目标。请认真完成下列步骤。

4.1　指导(1 小时 10 分钟)

Struts 2 局部类型转换

我们接下来解决这样的问题，例如，通过输入类似于“a,b”这样形式的数据，到后台自动封装成一个 Point 坐标类，这就需要我们将输入的 String 类型数据“a,b”自动转换成一个 Point 类的对象；当需要在页面上将 Point 类的对象信息显示出来，需要我们将 Point 类的对象再转换成 String 类型。

按照理论部分，我们应该编写一个类型转换类继承 ognl.DefaultTypeConverter，但是在这里，官方给我们提供了一个更合适的类型转换类 org.apache.struts2.util.StrutsTypeConverter，这个类也继承了 DefaultTypeConverter 类，只不过开发者一般更喜欢用 StrutsTypeConverter，接下来我们来分析 StrutsTypeConverter 应用示例。

第一步，编写 Point 类，Point 类的代码如示例代码 4-1 所示。

示例代码 4-1　编写 Point 类

```
package com.xtgj.model;
public class Point {
        private Integer x;
        private Integer y;
        public Integer getX() {
              return x;
```

```
    }
    public void setX(Integer x) {
        this.x = x;
    }
    public Integer getY() {
        return y;
    }
    public void setY(Integer y) {
        this.y = y;
    }
}
```

第二步，编写 PointConverter.java 类，代码如示例代码 4-2 所示。

示例代码 4-2 PointConverter 类

```
package com.xtgj.struts2.conversion;
import java.util.Map;
import org.apache.struts2.util.StrutsTypeConverter;
import com.xtgj.model.Point;
public class PointConverter extends StrutsTypeConverter {
    @Override
    public Object convertFromString(Map context, String[] values, Class toType) {
        Point p = new Point();
        String[] params = values[0].split(",");
        p.setX(Integer.valueOf(params[0]));
        p.setY(Integer.valueOf(params[1]));
        return p;
    }
    @Override
    public String convertToString(Map context, Object toType) {
        Point p = (Point)toType;
        StringBuffer buffer = new StringBuffer();
        buffer.append("Point[ x => ");
        buffer.append(p.getX());
        buffer.append(", y = > ");
        buffer.append(p.getY());
        buffer.append(" ]");
        return buffer.toString();
```

```
        }
    }
```

- convertFromString() 方法：说明从 String 类型转换成别的类型；
- convertToString() 方法：说明是从别的类型转换成 String 类型。

引入参数中的 toType 表示的是目标类型（是字符串类型还是 Point 类型），当客户端提交字符串请求时（即 toType==Point.class），values 指的是 textfield 中填写的字符串。对 if(Point.class==toType) 过程解释：

（1）String[] str=(String[])value;——将 value 显示转化为字符数组后赋值为数组 str[]。

（2）String[] paramValues=str[0].split(",");——取出 str[] 的第一个元素，即 str[0]，当 textfield 里填写的是“20,30”，str[0] 里的内容也就是一个字符串“20,30”，然后通过 str[0].split(",") 将字符串“20,30”以逗号为界限拆分成一个数组并赋值给 paramValues[]，这时 paramValues[] 里内容是 {“20”,“30”}。

（3）int x=Integer.parseInt (paramValues [0]);——将字符串类型的“20”转换为 integer 型赋值给 x。

第三步，编写 PointAction.java 类，代码如示例代码 4-3 所示。

示例代码 4-3　PointAction 类

```
package com.xtgj.action;
import com.xtgj.model.Point;
public class PointAction {
    private Point point;// 点对象成员
    public Point getPoint() {
        return point;
    }
    public void setPoint(Point point) {
        this.point = point;
    }
    public String execute() {
        System.out.println(this.point);
        return "success";
    }
}
```

第四步，配置属性文件使得让 Point 属性由 PointConverter 处理。

- 在需要类型转换的 action 所在的包下新建一个文件 PointAction-conversion.properties，其中：

PointAction：要转换的属性所在的类的类名；

-conversion.properies：固定格式。

● 填写属性文件内容：point=com.xtgj.struts2.conversion.PointConverter

point：具体要转换哪一个属性；

com.xtgj.struts2.conversion.PointConverter：具体用哪个 convert 类来转换 point。

```
point=com.xtgj.struts2.conversion.PointConverter
```

4.2 练习(50 分钟)

按照"4.1 指导"的提示，应用局部类型转换方式完成整个案例，即在输入页面输入"3,4"这样类似的数据（图 4-1）。

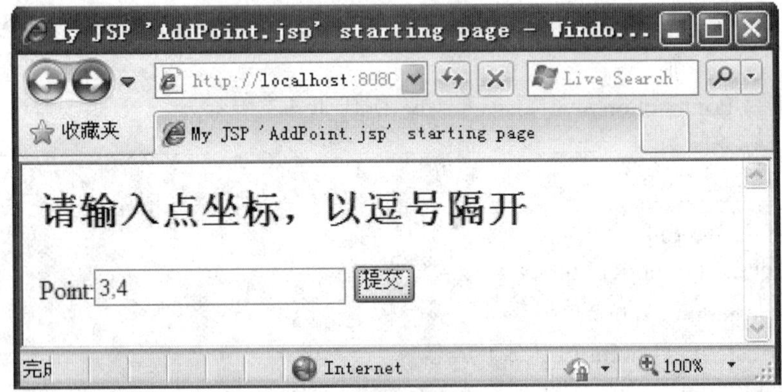

图 4-1 输入界面

输入界面 AddPoint.jsp 代码如示例代码 4-4 所示。

示例代码 4-4 AddPoint.jsp

```
<%@ page language="java" import="java.util.*" pageEncoding="UTF-8"%>
<!DOCTYPE HTML PUBLIC "-//W3C//DTD HTML 4.01 Transitional//EN">
<html>
    <head>
        <title>My JSP 'AddPoint.jsp' starting page</title>
    </head>
    <body>
        <h2>
            请输入点坐标，以逗号隔开
        </h2>
        <form action="point.action">
            Point:
```

```
                <input name="point" type="text" />
                <input name="submit" type="submit" value=" 提交 " />
            </form>
        </body>
    </html>
```

提交后，经过转换器转换后的结果如图 4-2 所示。

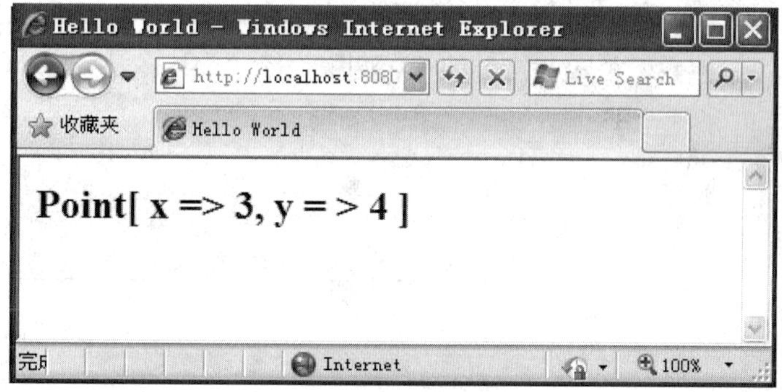

图 4-2　输出界面

显示 Point 类对象的信息的页面 ShowPoint.jsp 代码如示例代码 4-5 所示。

示例代码 4-5　ShowPoint.jsp

```
<%@ page contentType="text/html; charset=UTF-8"%>
<%@ taglib uri="/struts-tags" prefix="s"%>
<html>
    <head>
        <title>My JSP 'ShowPoint.jsp' starting page</title>
    </head>
    <body>
        <h2>
            <s:property value="point" />
        </h2>
    </body>
</html>
```

最后，struts.xml 中的配置如示例代码 4-6 所示。

示例代码 4-6　struts.xml 配置文件

```
<?xml version="1.0" encoding="UTF-8" ?>
<!-- 指定 Struts 2 配置文件的 DTD 信息 -->
<!DOCTYPE struts PUBLIC
```

```
        "-//Apache Software Foundation//DTD Struts Configuration 2.3//EN"
        "http://struts.apache.org/dtds/struts-2.3.dtd">
<!-- struts 是 Struts 2 配置文件的根元素 -->
<struts>
    <constant name="struts.custom.i18n.resources"
        value="globalMessages" />
    <package name="ConverterDemo" extends="struts-default">
        <action name="point" class="com.xtgj.action.PointAction">
            <result>/ShowPoint.jsp</result>
            <result name="input">/AddPoint.jsp</result>
        </action>
    </package>
</struts>
```

4.3 作业

将“4.2 练习”改为采用全局数据类型转换方式来实现。

第 5 章 Struts 2 表单数据校验

本阶段目标

完成本章内容后，你将能够学到：

✧ 数据校验的主要原理。

✧ 数据校验的组成部分。

✧ 数据校验的使用语法。

本阶段给出的步骤全面详细，请读者按照给出的上机步骤独立完成上机练习，以达到要求的学习目标。请认真完成下列步骤。

5.1 指导(1 小时 10 分钟)

Struts 2 数据校验

系统提供的常用的校验器如下：

- required（必填校验器，要求 field 的值不能为 null）；
- requiredstring（必填字符串校验器，要求 field 的值不能为 null，并且长度大于 0，默认情况下会对字符串去前后空格）；
- stringlength（字符串长度校验器，要求 field 的值必须在指定的范围内，否则校验失败，minLength 参数指定最小长度，maxLength 参数指定最大长度，trim 参数指定校验 field 之前是否去除字符串前后的空格）；
- regex（正则表达式校验器，检查被校验的 field 是否匹配一个正则表达式 .expression 参数指定正则表达式，caseSensitive 参数指定进行正则表达式匹配时，是否区分大小写，默认值为 true）；
- int（整数校验器，要求 field 的整数值必须在指定范围内，min 指定最小值，max 指定最大值）；
- double（双精度浮点数校验器，要求 field 的双精度浮点数必须在指定范围内，min 指定最小值，max 指定最大值）；
- fieldexpression（字段 ognl 表达式校验器，要求 field 满足一个 ognl 表达式，expression 参数指定 ognl 表达式，该逻辑表达式基于 ValueStack 进行求值，返回 true 时校验通过，否则不通过）；

- email（邮件地址校验器，要求如果 field 的值非空，则必须是合法的邮件地址）；
- url（网址校验器，要求如果 field 的值非空，则必须是合法的 url 地址）；
- date（日期校验器，要求 field 的日期值必须在指定范围内，min 指定最小值 ,max 指定最大值）；
- conversion（转换校验器，指定在类型转换失败时，提示的错误信息）；
- visitor（用于校验 action 中的复合属性，它指定一个校验文件用于校验复合属性中的属性）；
- expression（ognl 表达式校验器，expression 参数指定 ognl 表达式，该逻辑表达式基于 ValueStack 进行求值，返回 true 时校验通过，否则不通过，该校验器不可用在字段校验器风格的配置中）。

校验器使用举例：

（1）required：必填校验器。

```
<field-validator type="required">
        <message> 性别不能为空 !</message>
</field-validator>
```

（2）requiredstring：必填字符串校验器。

```
<field-validator type="requiredstring">
        <param name="trim">true</param>
        <message> 用户名不能为空 !</message>
</field-validator>
```

（3）stringlength：字符串长度校验器 。

```
<field-validator type="stringlength">
        <param name="maxLength">10</param>
        <param name="minLength">2</param>
        <param name="trim">true</param>
        <message><![CDATA[ 产品名称应在 2-10 个字符之间 ]]></message>
</field-validator>
```

（4）email：邮件地址校验器。

```
<field-validator type="email">
            <message> 电子邮件地址无效 </message>
</field-validator>
```

（5）regex：正则表达式校验器。

```
<field-validator type="regex">
        <param name="expression"><![CDATA[^1[358]\d{9}$]]></param>
        <message> 手机号格式不正确 !</message>
</field-validator>
```

（6）int：整数校验器。

```
<field-validator type="int">
        <param name="min">1</param>
        <param name="max">150</param>
        <message> 年龄必须在 1-150 之间 </message>
</field-validator>
```

（7）fieldexpression：字段 ognl 表达式校验器。

```
<field name="imagefile">
    <field-validator type="fieldexpression">
        <param name="expression"><![CDATA[imagefile.length() <= 0]]></param>
        <message> 文件不能为空 </message>
    </field-validator>
</field>
```

5.2　练习（50 分钟）

按照“5.1 指导”的提示，利用 XML 配置方式实现对指定 action() 方法实现输入校验。本例实现校验用户注册的输入信息是否合法。

第一步，编写 UserAction.java 类，代码如示例代码 5-1 所示。

示例代码 5-1　编写 UserAction.java 类

```
package com.xtgj.struts2.chapter05.user;
import com.opensymphony.xwork2.ActionSupport;
public class UserAction extends ActionSupport {
        private String username;
        private String mobile;
        private static final long serialVersionUID = -6328095689264546407L;
```

```
        public String update() {
            return "success";
        }
        public String delete() {
            return "success";
        }
        public String login() {
            return "success";
        }
        public String regist() {
            return "success";
        }
        public String getUsername() {
            return username;
        }
        public void setUsername(String username) {
            this.username = username;
        }
        public String getMobile() {
            return mobile;
        }
        public void setMobile(String mobile) {
            this.mobile = mobile;
        }
    }
```

第二步，在 com.xtgj.struts2.chapter05.user 包下新建两个验证文件，UserAction-user_regist-validation.xml，此验证文件对应“user_regist”请求，只对 regist() 方法验证，其代码如示例代码 5-2 所示。

示例代码 5-2 UserAction-user_regist-validation.xml

```
<?xml version="1.0" encoding="UTF-8"?>
<!DOCTYPE validators PUBLIC
"-//OpenSymphony Group//XWork Validator 1.0.3//EN"
"http://www.opensymphony.com/xwork/xwork-validator-1.0.3.dtd">
<validators>
    <field name="username">
        <field-validator type="requiredstring">
```

```
                <param name="trim">true</param>
                <message> 用户名不能为空 !</message>
            </field-validator>
        </field>
        <field name="mobile">
            <field-validator type="requiredstring">
                <param name="trim">true</param>
                <message> 电话号码不能为空 !</message>
            </field-validator>
            <field-validator type="regex">
                <param name="expression"><![CDATA[^1[358]\\d{9}]]></param>
                <message> 电话号码格式不正确 </message>
            </field-validator>
        </field>
</validators>
```

UserAction-user_update-validation.xml，此验证文件对应“user_update”请求，只对 update() 方法验证，其代码如示例代码 5-3 所示。

示例代码 5-3　UserAction-user_update-validation.xml

```
<?xml version="1.0" encoding="UTF-8"?>
<!DOCTYPE validators PUBLIC
"-//OpenSymphony Group//XWork Validator 1.0.3//EN"
"http://www.opensymphony.com/xwork/xwork-validator-1.0.3.dtd">
<validators>
    <field name="username">
        <field-validator type="requiredstring">
            <param name="trim">true</param>
            <message> 用户名不能为空 !</message>
        </field-validator>
    </field>
        <field name="mobile">
        <field-validator type="requiredstring">
            <param name="trim">true</param>
            <message> 电话号码不能为空 !</message>
        </field-validator>
        <field-validator type="regex">
            <param name="expression"><![CDATA[^1[358]\\d{9}]]></param>
            <message> 电话号码格式不正确 </message>
```

```
            </field-validator>
        </field>
    </validators>
```

第三步，编写用户注册页面，代码请参考理论部分 register.jsp。

第四步，配置 user.xml 文件，其代码如示例代码 5-4 所示。

示例代码 5-4 配置 user.xml 文件

```
<?xml version="1.0" encoding="UTF-8"?>
<!DOCTYPE struts PUBLIC
    "-//Apache Software Foundation//DTD Struts Configuration 2.3//EN"
    "http://struts.apache.org/dtds/struts-2.3.dtd">
<struts>
    <package name="user" namespace="/user" extends="struts-default">
        <global-results>
            <result name="msg">msg.jsp</result>
        </global-results>
        <action  name="user_*"  class="com.xtgj.struts2.chapter05.user.UserAction"
method="{1}">
            <result name="success">success.jsp</result>
            <result name="input">register.jsp</result>
        </action>
    </package>
</struts>
```

第五步，配置 struts.xml，其代码请参考理论部分。

至此用户注册的校验功能就完成了，请读者根据上述练习设计用户信息修改业务的数据校验功能，并实战。

5.3 作业

根据 5.2 的练习设计用户信息表，完成注册和信息修改业务的 DAO 实现，同时实现数据校验功能。

第 6 章　Struts 2 拦截器

本阶段目标

完成本章内容后，你将能够学到：

✧ 拦截器主要原理。

✧ 自定义拦截器。

本阶段给出的步骤全面详细，请读者按照给出的上机步骤独立完成上机练习，以达到要求的学习目标。请认真完成下列步骤。

6.1　指导(1 小时 10 分钟)

Struts 2 自定义拦截器

在实际开发中，我们经常会遇到这样的问题，例如，对于部门信息增、删、改的动作需要有管理员的权限才能够进行，普通用户只能够查看，若这种操作全部由后台代码实现，是非常繁琐的。在 Struts 2 中，这个问题可以通过给特定的 action 指定拦截器来解决。接下来我们来分析一个拦截器的应用示例，该示例实现对部门信息增、删、改的动作进行拦截并验证是否有执行此动作的权限，实现的效果如图 6-1 所示。

首先，用户直接访问部门信息列表页面，此时不需要用户的管理员权限。

图 6-1　部门列表

接着，在没有管理员权限验证的情况下点击部门信息列表里的“删除”链接，此时由于权限不足，将会出现如图 6-2 所示的消息反馈界面。

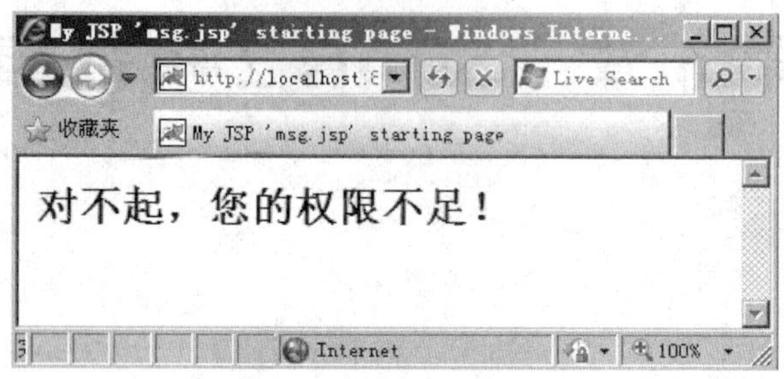

图 6-2　消息反馈界面

同理，若在部门信息列表页面上点击“添加部门”链接，会显示如图 6-3 所示的部门信息添加页面，但是当用户填入信息点击“提交”按钮时，仍然会看到如图 6-2 所示的信息反馈界面。修改的操作也是一样的。

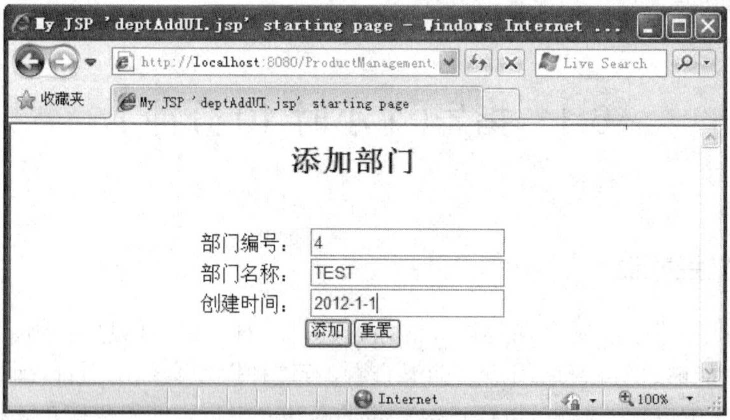

图 6-3　部门添加

假设用户以管理员身份登录系统，如图 6-4 所示，仍然访问部门信息列表页面，此时该用户已经通过了管理员权限验证，此时，点击部门信息列表里的“删除”链接，将会删除成功，如图 6-5 所示。

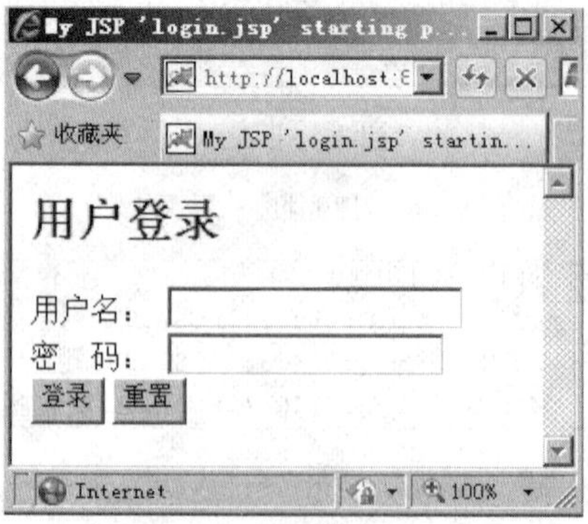

图 6-4　管理员登录页面

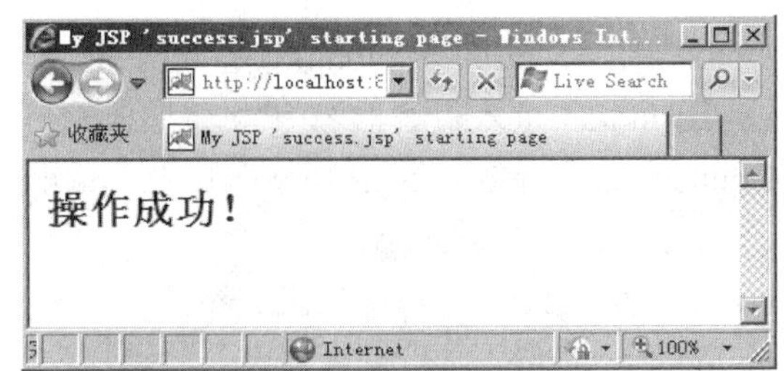

图 6-5　操作成功页面

部门信息添加和修改的操作同上，这里不再赘述。

6.2　练习（50 分钟）

按照"6.1 指导"的提示，应用自定义拦截器方式完成整个案例。

第一步，定义拦截器类 PermissionInterceptor，其代码如示例代码 6-1 所示。

示例代码 6-1　定义拦截器类 PermissionInterceptor

```
package com.xtgj.interceptor;
import java.util.Map;
import com.opensymphony.xwork2.ActionContext;
import com.opensymphony.xwork2.ActionInvocation;
import com.opensymphony.xwork2.interceptor.Interceptor;
public class PermissionInterceptor implements Interceptor {
	private static final long serialVersionUID = -5178310397732210602L;
	private String ignoreActions;
	public String getIgnoreActios() {
		return ignoreActions;
	}
	public void setIgnoreActions(String ignoreActions) {
		this.ignoreActions = ignoreActions;
	}
	public void destroy() {
	}
	public void init() {
	}
```

```
        public String intercept(ActionInvocation invocation) throws Exception {
            System.out.println(" 进入拦截器……");
            ActionContext ctx = invocation.getInvocationContext();
            Map session = ctx.getSession();
            String user = (String) session.get("username");
            boolean ignore = false;
            String currentAction = invocation.getProxy().getActionName();
            String[] actions = ignoreActions.split(",");
            for (String action : actions) {
                if (currentAction.matches(action.trim())) {
                    ignore = true;
                    break;
                }
            }
            if (user != null || ignore == true) {
                String result = invocation.invoke();
                return result;
            } else {
                ctx.put("msg", " 对不起,您的权限不足！ ");
                return "msg";
            }
        }
    }
```

PermissionInterceptor 实现了 com.opensymphony.xwork2.interceptor.Interceptor 接口，覆盖了 intercept() 方法。在方法中首先获取登录用户信息，若登录信息不为空，则援引方法用 action 中的方法，即通过验证，继续 action 中的任务；若登录信息为空，则拦截该动作并返回 name 属性为“msg”的 result 视图。这个拦截其中还定义了一个成员变量 ignoreActions，这个属性的值是由 struts.xml 中的 Interceptor 配置注入的，代表可以忽略拦截器的所有 action 名称（以“,”隔开）。如果登录信息为空，但是当前请求 action 的名称正好和 ignoreActions 中的某个 Action 名称匹配，则仍然忽略拦截器。

第二步，定义域类 Dept，代码如示例代码 6-2 所示。

示例代码 6-2　Dept 类

```
package com.xtgj.domain;
import java.util.Date;
public class Dept {
	private Integer deptno;
	private String dname;
	private Date creationTime;
	// 省略 getter/setter
	// 省略构造器
}
```

第三步，定义 DeptAction.java 类，代码如示例代码 6-3 所示。

示例代码 6-3　DeptAction 类

```
package com.xtgj.dept;
import java.util.ArrayList;
import java.util.Date;
import java.util.List;
import com.xtgj.domain.Dept;
public class DeptAction {
	private Dept dept;
	private List<Dept> list = new ArrayList<Dept>();;
	// 省略 getter/setter
	public String list() {
		list.add(new Dept(1, "SALE", new Date()));
		list.add(new Dept(2, "CLERK", new Date()));
		list.add(new Dept(3, "MANAGE", new Date()));
		return "list";
	}
	public String add() {
		return "success";
	}
	public String del() {
		return "success";
	}
	public String updateUI() {
		return "updateUI";
	}
```

```
    public String addUI() {
        return "addUI";
    }
}
```

在这个类中，只是简单地实现了业务的模拟流程，没有具体实现 DAO 操作，请读者在本例基础之上实现真正的业务，并实战。

第四步，配置 struts.xml 文件，代码如示例代码 6-4 所示。

示例代码 6-4　struts.xml 文件

```
<?xml version="1.0" encoding="UTF-8"?>
<!DOCTYPE struts PUBLIC
    "-//Apache Software Foundation//DTD Struts Configuration 2.3//EN"
    "http://struts.apache.org/dtds/struts-2.3.dtd">
<struts>
    <package name="dept" namespace="/dept" extends="struts-default">
        <interceptors>
            <interceptor name="permission"
                class="com.xtgj.interceptor.PermissionInterceptor">
                <param name="ignoreActions">dept_list,dept_addUI</param>
            </interceptor>
            <interceptor-stack name="permissionStack">
                <interceptor-ref name="defaultStack" />
                <interceptor-ref name="permission" />
            </interceptor-stack>
        </interceptors>
        <global-results>
            <result name="msg">msg.jsp</result>
        </global-results>
        <action name="dept_*" class="com.xtgj.dept.DeptAction"
            method="{1}">
            <result name="success">success.jsp</result>
            <result name="list">deptList.jsp</result>
            <result name="updateUI">deptUpdateUI.jsp</result>
            <result name="addUI">deptAddUI.jsp</result>
            <interceptor-ref name="permissionStack" />
        </action>
    </package>
</struts>
```

第五步，在 WebRoot 目录下创建 dept 文件夹，在 dept 文件夹中创建视图 deptList.jsp，代码如示例代码 6-5 所示。

示例代码 6-5　deptList.jsp

```
<%@ page language="java" import="java.util.*" pageEncoding="UTF-8"%>
<%@taglib prefix="c" uri="http://java.sun.com/jsp/jstl/core"%>
<!DOCTYPE HTML PUBLIC "-//W3C//DTD HTML 4.01 Transitional//EN">
<html>
    <head>
        <title>My JSP 'deptList.jsp' starting page</title>
    </head>

    <body>
        <center>
            <h2>
                部门列表
            </h2>
            <a href="dept_addUI"> 增加部门 </a>
        </center>
        <br>
        <table align="center" border="1">
            <tr>
                <td>
                    部门号
                </td>
                <td>
                    部门名
                </td>
                <td>
                    部门创建时间
                </td>
                <td colspan="2">
                    操作
                </td>
            </tr>
            <c:forEach items="${list}" var="dept">
                <tr>
                    <td>
```

```
                        <c:out value="${dept.deptno}" />
                    </td>
                    <td>
                        <c:out value="${dept.dname}" />
                    </td>
                    <td>
                        <c:out value="${dept.creationTime}" />
                    </td>
                    <td>
                        <a   href="dept_del.action?dept.deptno=${dept.deptno}">
删除 </a>
                    </td>
                    <td>
                        <a   href="dept_updateUI?dept.deptno=${dept.deptno}">
修改 </a>
                    </td>
                </tr>
            </c:forEach>
        </table>
    </body>
</html>
```

第六步，创建添加部门页面，修改部门页面，具体内容参照前面的章节（第 3 章）中的代码。

第七步，创建操作失败的视图页面 msg.jsp，代码如示例代码 6-6 所示。

示例代码 6-6　msg.jsp

```
<%@ page language="java" import="java.util.*" pageEncoding="UTF-8"%>
<!DOCTYPE HTML PUBLIC "-//W3C//DTD HTML 4.01 Transitional//EN">
<html>
    <head>
        <title>My JSP 'msg.jsp' starting page</title>
    </head>
    <body>
        <h2>${msg }</h2>
    </body>
</html>
```

我们假设用户已经登录，此时 session 中已经存储了用户的登录信息，为了简化本例操作，我们直接将 PermissionInterceptor 中的用户信息赋予特定值，如示例代码 6-7 中标为粗体的部分。

示例代码 6-7　存储用户信息

```
// 省略部分代码
public String intercept(ActionInvocation invocation) throws Exception {
	System.out.println(" 进入拦截器……");
	String user = "Admin";
	boolean ignore = false;
	String currentAction = invocation.getProxy().getActionName();
	String[] actions = ignoreActions.split(",");
	for (String action : actions) {
		if (currentAction.matches(action.trim())) {
			ignore = true;
			break;
		}
	}
	if (user != null || ignore == true) {
		String result = invocation.invoke();
		return result;
	} else {
		ctx.put("msg", " 对不起，您的权限不足！ ");
		return "msg";
	}
}
```

第八步，创建操作成功的视图页面 success.jsp，如示例代码 6-8 所示。

示例代码 6-8　success.jsp

```
<%@ page language="java" import="java.util.*" pageEncoding="UTF-8"%>
<!DOCTYPE HTML PUBLIC "-//W3C//DTD HTML 4.01 Transitional//EN">
<html>
	<head>
		<title>My JSP 'success.jsp' starting page</title>
	</head>
	<body>
		<h2> 操作成功！ </h2>
	</body>
</html>
```

至此,利用自定义拦截器实现部门管理模块的权限验证功能就完成了。

6.3 作业

根据部门管理的业务需求分析,创建部门表 dept,如表 6-1 所示,创建管理员表 admin,如表 6-2 所示。实现部门 CRUD 操作,要求部门的 CRU 操作必须通过权限验证后才能够进行,利用自定义拦截器实现。

表 6-1 dept 表

字段名	类型	是否空	是否主键
deptno	int	否	是
dname	varchar(20)	否	否
creationTime	datetime	否	否

表 6-2 admin 表

字段名	类型	是否空	是否主键
adminId	int	否	是
adminName	varchar(20)	否	否
adminPassword	varchar(20)	否	否

第 7 章　Struts 2 标签库

本阶段目标

完成本章内容后，你将能够学到：

✧ OGNL 表达式的运用。

✧ Struts 2 中的表单标签。

本阶段给出的步骤全面详细，请读者按照给出的上机步骤独立完成上机练习，以达到要求的学习目标。请认真完成下列步骤。

7.1　指导(1 小时 10 分钟)

在进行 Struts 2 的标签库介绍之前，有必要着重对 OGNL（Object Graph Navigating Language）对象导航语言做详细解析。因为在之后的演示代码中经常会用到一些有关 OGNL 的代码，为了不让读者一头雾水，也为了更好地学习 Struts 2 的标签库，因此，把 OGNL 当作学习 Struts 2 标签库的基础知识来介绍。让大家把 Struts 2 的基础打得更加扎实。

7.1.1　OGNL 表达式

OGNL 可以理解为：对象图形化导航语言。是一种可以方便地操作对象属性的开源表达式语言。OGNL 有如下特点：

- 支持对象方法调用，形式如 objName.methodName()。
- 支持类静态的方法调用和值访问，表达式的格式为 @[类全名（包括包路）]@[方法名 | 值名]，例如，@java.lang.String@format('foo %s', 'bar') 或 @tutorial.MyConstant@APP_NAME。
- 支持赋值操作和表达式串联，例如，price=100, discount=0.8, calculatePrice()，这个表达式会返回 80。
- 访问 OGNL 上下文 OGNL Context 和 ActionContext。
- 操作集合对象。

OGNL 有一个上下文（Context）概念，所谓上下文就是一个 MAP 结构，它实现了 java.utils.Map 接口，在 Struts 2 中上下文（Context）的实现为 ActionContext，图 7-1 是上下文（Context）的结构示意图。

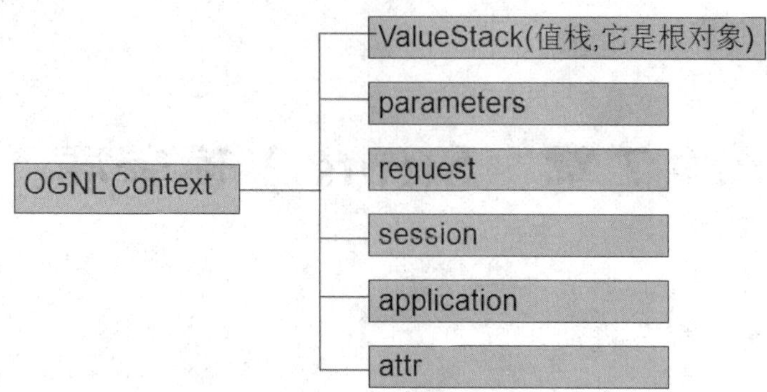

图 7-1 Struts 2 上下文结构示意图

当 Struts 2 接受一个请求时，会迅速创建 ActionContext、ValueStack、action 。然后把 action 存放进 ValueStack，所以 action 的实例变量可以被 OGNL 访问。OGNL 要结合 Struts 2 标签来使用。最常见的是“%”“#”“$”这三个符号的使用。由于 ValueStack（值栈）是 Struts 2 中 OGNL 的根对象，如果用户需要访问值栈中的对象，在 JSP 页面可以直接通过下面的 EL 表达式访问 ValueStack（值栈）中对象的属性，例如，通过 ${msg} 表达式可以获得值栈中某个对象的 msg 属性。$ 广泛应用于 EL 中，比较容易理解，这里重点介绍“#”和“%”符号的用法。

1. “#”符号的三种用途

（1）访问非根对象（Struts 2 中值栈为根对象）

如 OGNL 上下文和 action 上下文，“#”相当于 ActionContext.getContext()，如果访问其他 Context 中的对象，由于它们不是根对象，所以在访问时，需要添加“#”前缀。以下是几个 ActionContext 中有用的属性：

- application 对象：用于访问 ServletContext，例如 #application.userName 或者 #application['userName']，相当于调用 ServletContext 的 getAttribute("username")；
- session 对象：用来访问 HttpSession，例如 #session.userName 或者 #session['userName']，相当于调用 session.getAttribute("userName")；
- request 对象：用来访问 HttpServletRequest 属性（attribute）的 Map，例如 #request.userName 或者 #request['userName']，相当于调用 request.getAttribute("userName")；
- parameters 对象：用于访问 HTTP 的请求参数，例如 #parameters.userName 或者 #parameters ['user Name']，相当于调用 request.getParameter("username")。
- attr 对象：用于按 page->request->session->application 顺序访问其属性。

（2）创建 List/Map 集合对象

创建 List 集合对象，如示例代码 7-1 所示。

示例代码 7-1 利用标签生成 List 对象

```
<s:set name="list" value="{'zhangming','xiaoi','liming'}" />
<s:iterator value="#list" id="n">
      <s:property value="n"/><br>
</s:iterator>
```

生成 Map 对象。创建 Map 对象，如示例代码 7-2 所示。

示例代码 7-2　利用标签生成 Map 对象

```
<s:set name="foobar" value="#{'foo1':'bar1', 'foo2':'bar2'}" />
<s:iterator value="#foobar" >
    <s:property value="key"/>=<s:property value="value"/><br>
</s:iterator>
```

set 标签用于将某个值放入指定范围。其中 scope 属性用于指定变量被放置的范围，该属性可以接受 application、session、request、page 或 action。如果没有设置该属性，则默认放置在 OGNL Context 中。Value 属性赋予变量的值，如果没有设置该属性，则将 ValueStack 栈顶的值赋予变量。

对于集合类型，OGNL 表达式可以使用 in 和 not in 两个元素符号。其中，in 表达式用来判断某个元素是否在指定的集合对象中。not in 判断某个元素是否在指定的集合对象中，如示例代码 7-3 所示。

in 表达式：

示例代码 7-3　in 表达式

```
<s:if test="'foo' in {'foo','bar'}">
        在
</s:if>
<s:else>
        不在
</s:else>
```

not in 表达式如示例代码 7-4 所示。

示例代码 7-4　not in 表达式

```
<s:if test="'foo' not in {'foo','bar'}">
          不在
</s:if>
<s:else>
          在
</s:else>
```

（3）用于过滤和投影（projecting）集合

除了 in 和 not in 之外，OGNL 还允许使用某个规则获得集合对象的子集，常用的有以下 3 个相关操作符。

- ?：获得所有符合逻辑的元素。
- ^：获得符合逻辑的第一个元素。

● $:获得符合逻辑的最后一个元素。

如示例代码 7-5 所示。

示例代码 7-5 过滤和投影集合示例

```
<s:iterator value="books.{?#this.price > 35}">
    <s:property value="title" /> - $<s:property value="price" /><br>
</s:iterator>
```

在上述代码中,直接在集合后紧跟 .{} 运算符表明用于取出该集合的子集,.{} 内的表达式用于获取符合条件的元素,this 指的是为了从大集合 books 筛选数据到小集合,需要对大集合 books 进行迭代,this 代表当前迭代的元素。本例的表达式用于获取集合中价格大于 35 的书集合。如示例代码 7-6 所示。

示例代码 7-6 获取集合中价格大于 35 的书集合

```
public class BookAction {
    private List<Book> books;
    // 省略部分代码……
    public String execute() {
        books = new LinkedList<Book>();
        books.add(new Book("A735619678", "spring", 67));
        books.add(new Book("B435555322", "ejb3.0",15));
    }
}
```

2. "%"符号的用途

在标签的属性值被理解为字符串类型时,告诉执行环境 %{} 里的是 OGNL 表达式。实际上就是让被理解为字符串的表达式,被真正当成 OGNL 来执行。这一点类似 JavaScript 里面的 eval_r () 方法的功能,例如:

```
var oDiv = eval_r("document.all.div"+index);
```

当 index 变量为 1 时,语句就会被当作下面的代码来执行。

```
var oDiv = document.all.div1;
```

%{} 就是起这个作用。如示例代码 7-7 所示。

示例代码 7-7 %{} 的作用

```
<s:set name="myMap" value="#{'key1':'value1','key2':'value2'}"/>
<s:property value="#myMap['key1']"/>
<s:url value="#myMap['key1']" />
```

上述代码的第 2 行会在页面上输出“value1”，而第 3 行则会输出 "#myMap['key1']" 的字符串。如果将第 3 行改写为：

```
<s:url value="%{#myMap['key1']}"/>
```

输出为“value1”。这说明 struts 2 中不同的标签对 OGNL 的表达式的理解是不一样的。如果当有的标签“看不懂”类似“#myMap['key1']”的语句时，就要用 %{} 来标示相关语句进行“翻译”。

3. “$”有两种特殊的用途

（1）在国际化资源文件中，引用 OGNL 表达式

例如，有一个字段 x，需要被验证为指定上限和下限范围之内的整型变量，我们可以将国际化资源文件中的代码设定如下：

```
reg.agerange= 国际化资源信息：年龄必须在 ${min} 同 ${max} 之间。
```

在 Struts 2 框架的配置文件中引用 OGNL 表达式，如示例代码 7-8 所示。

示例代码 7-8　在 Struts 2 框架的配置文件中引用 OGNL 表达式

```
<validators>
        <field name="intb">
            <field-validator type="int">
            <param name="min">10</param>
            <param name="max">100</param>
            <message>BAction-test 校验：数字必须为 ${min} 为 ${max} 之间！</
message>
          </field-validator>
        </field>
</validators>
```

（2）在 Struts 2 配置文件中，引用 OGNL 表达式

如示例代码 7-9 所示。

示例代码 7-9　在 Struts 2 配置文件中引用 OGNL 表达式

```
<action name="saveUser" class="userAction" method="save">
    <result type="redirect">listUser.action?msg=${msg}</result>
</action>
```

7.1.2　Struts 2 表单标签

Struts 2 中的表单标签主要包括：<s:form/>、<s:textfield/>、<s:select/>、< s:radio/>、<s:checkboxlist/> 等，这里我们介绍其中较为常用的几个。

（1）表单标签 _checkboxlist 复选框

如果集合为 list，则代码片段如示例代码 7-10 所示。

示例代码 7-10 用 list 集合表示复选框

```
<s:checkboxlist name="list" list="{'Java','.Net','RoR','PHP'}" value="{'Java','.Net'}"/>
```

生成如示例代码 7-11 所示的 html 代码。

示例代码 7-11 用 list 集合表示复选框所生成的 html 代码

```
<input type="checkbox" name="list" value="Java" checked="checked"/><label>Java</label>
<input type="checkbox" name="list" value=".Net" checked="checked"/><label>.Net</label>
<input type="checkbox" name="list" value="RoR"/><label>RoR</label>
<input type="checkbox" name="list" value="PHP"/><label>PHP</label>
```

如果集合为 map，则代码片段如示例代码 7-12 所示。

示例代码 7-12 用 map 集合表示复选框

```
<s:checkboxlist name="map" list="#{1:' 瑜珈用品 ',2:' 户外用品 ',3:' 球类 ',4:' 自行车 '}" listKey="key" listValue="value" value="{1,2,3}"/>
```

生成如示例代码 7-15 所示的 html 代码。

示例代码 7-13 用 map 集合表示复选框所生成的 html 代码

```
<input type="checkbox" name="map" value="1" checked="checked"/><label> 瑜珈用品 </label>
<input type="checkbox" name="map" value="2" checked="checked"/><label> 户外用品 </label>
<input type="checkbox" name="map" value="3" checked="checked"/><label> 球类 </label>
<input type="checkbox" name="map" value="4"/><label> 自行车 </label>
```

如果集合里存放的是 JavaBean，则代码片段如示例代码 7-14 所示。

示例代码 7-14 集合中存放的是 JavaBean

```
<%
  Person person1 = new Person(1," 第一个 ");
  Person person2 = new Person(2," 第二个 ");
  List<Person> list = new ArrayList<Person>();
```

```
        list.add(person1);
        list.add(person2);
        request.setAttribute("persons",list);
        %>
    <s:checkboxlist     name="beans"     list="#request.persons"     listKey="personid"
listValue="name"/>
```

生成如示例代码 7-15 所示 html 代码。

示例代码 7-15　集合中存放的是 JavaBean 生成的 html 代码

```
<input type="checkbox" name="beans" value="1"/><label> 第一个 </label>
<input type="checkbox" name="beans" value="2"/><label> 第二个 </label>
```

（2）表单标签 _radio 单选框

该标签的使用和 checkboxlist 复选框相同。如果集合里存放的是 JavaBean，则代码片段如示例代码 7-16 所示。

示例代码 7-16　表单标签 _radio 单选框

```
< s:radio name="beans" list="#request.persons" listKey="personid" listValue="name"/>
```

生成如示例代码 7-17 所示的 html 代码。

示例代码 7-17　表单标签 _radio 单选框生成如下 html 代码

```
<input type="radio" name="beans" id="beans1" value="1"/><label> 第一个 </label>
<input type="radio" name="beans" id="beans2" value="2"/><label> 第二个 </label>
```

如果集合为 map，则代码片段如示例代码 7-18 所示。

示例代码 7-18　用 map 集合表示单选框

```
<s:radio name="map" list="#{1:' 瑜珈用品 ',2:' 户外用品 ',3:' 球类 ',4:' 自行车 '}" list-
Key="key" listValue="value" value="1"/>
```

生成如示例代码 7-19 所示的 html 代码。

示例代码 7-19　用 map 集合表示单选框生成如下 html 代码

```
<input type="radio" name="map" id="map1" value="1"/><label for="map1"> 瑜珈用
品 </label>
<input type="radio" name="map" id="map2" value="2"/><label for="map2"> 户外用
品 </label>
```

```
<input type="radio" name="map" id="map3" value="3"/><label for="map3"> 球类 </label>
<input type="radio" name="map" id="map4" value="4"/><label for="map4"> 自行车 </label>
```

如果集合为 list，则代码片段如示例代码 7-20 所示。

示例代码 7-20 用 list 集合表示单选框

```
<s:radio name="list" list="{'Java','.Net'}" value="'Java'"/>
```

生成示例代码 7-21 所示的 html 代码。

示例代码 7-21 用 list 集合表示单选框生成如下 html 代码

```
<input type="radio" name="list" checked="checked" value="Java"/><label>Java</label>
<input type="radio" name="list" value=".Net"/><label>.Net</label>
```

（3）表单标签 _select 下拉选择框

如果集合为 list，则代码片段如示例代码 7-22 所示。

示例代码 7-22 在表单标签中用 list 集合表示下拉选择框

```
<s:select name="list" list="{'Java','.Net'}" value="'Java'"/>
```

生成如示例代码 7-23 所示的 html 代码。

示例代码 7-23 在表单标签中用 list 集合表示下拉选择框生成如下 html 代码

```
<select name="list" id="list">
    <option value="Java" selected="selected">Java</option>
    <option value=".Net">.Net</option>
</select>
```

如果集合里存放的是 JavaBean，则代码片段如示例代码 7-24 所示。

示例代码 7-24 list 集合里存放的是 JavaBean

```
<s:select name="beans" list="#request.persons" listKey="personid" listValue="name"/>
```

生成如示例代码 7-25 所示的 html 代码。

示例代码 7-25　list 集合里存放的是 JavaBean 生成如下 html 代码

```
<select name="beans" id="beans">
    <option value="1"> 第一个 </option>
    <option value="2"> 第二个 </option>
</select>
```

如果集合为 map，则代码片段如示例代码 7-26 所示。

示例代码 7-26　map 集合里存放的是 JavaBean

```
<s:select name="map" list="#{1:' 瑜珈用品 ',2:' 户外用品 ',3:' 球类 ',4:' 自行车 '}"
listKey="key" listValue="value" value="1"/>
```

生成如示例代码 7-27 所示的 html 代码。

示例代码 7-27　map 集合里存放的是 JavaBean 生成如下 html 代码

```
<select name="map" id="map">
    <option value="1" selected="selected"> 瑜珈用品 </option>
    <option value="2"> 户外用品 </option>
    <option value="3"> 球类 </option>
    <option value="4"> 自行车 </option>
</select>
```

（4）<s:token /> 标签防止重复提交

<s:token /> 标签防止重复提交，用法如下：

第一步，在表单中加入 <s:token />，如示例代码 7-28 所示。

示例代码 7-28　<s:token /> 标签防止重复提交

```
<s:form action="helloworld_other" method="post" namespace="/test">
  <s:textfield name="person.name"/><s:token/><s:submit/>
</s:form>
```

第二步，配置 action，如示例代码 7-29 所示。

示例代码 7-29　配置 Action

```
<action name="helloworld_*" class="com.xtgj.action.HelloWorldAction" method="{1}">
        <interceptor-ref name="defaultStack" />
         <interceptor-ref name="token" />
         <result name="invalid.token">/WEB-INF/page/message.jsp</result>
         <result>/WEB-INF/page/result.jsp</result>
</action>
```

以上配置加入了“token”拦截器和“invalid.token”结果，因为“token”拦截器在会话的 token 与请求的 token 不一致时，将会直接返回“invalid.token”结果。在 debug 状态，控制台出现报错信息，是因为 Action 中并没有 struts.token 和 struts.token.name 属性，我们不用关心这个错误。

7.2 练习(50 分钟)

按照“7.1 指导”的提示，给出一个完整的 Struts 2 标签和 OGNL 表达式相结合的示例，OGNL.jsp 代码如示例代码 7-30 所示。

示例代码 7-30 OGNL.jsp

```
<%@ page language="java" import="java.util.*" pageEncoding="UTF-8"%>
<%@page import="com.xtgj.struts2.domain.Sex"%>
<%@ taglib prefix="s" uri="/struts-tags"%>
<!DOCTYPE HTML PUBLIC "-//W3C//DTD HTML 4.01 Transitional//EN">
<html>
    <head>
        <title>My JSP 'OGNL.jsp' starting page</title>
    </head>
    <body>
      <%
                request.setAttribute("req", "request scope");
            request.getSession().setAttribute("sess", "session scope");
            request.getSession().getServletContext().setAttribute("app",
                    "aplication scope");
        %>
        1. 通过 ognl 表达式获取 属性范围中的值
        <br>
        <s:property value="#request.req" />
        <br />
        <s:property value="#session.sess" />
        <br />
        <s:property value="#application.app" />
        <br />
        <hr>
        2. 通过
```

```
<SPAN style="BACKGROUND-COLOR: #fafafa">ognl 表达式创建 list 集合，并且遍历出集合中的值 <br> <s:set name="list"
        value="{'eeeee','ddddd','ccccc','bbbbb','aaaaa'}"></s:set> <s:iterator
        value="#list" var="o">
        <!-- ${o }<br/> -->
        <s:property />
        <br />
    </s:iterator> <br />
    <hr>
3. 通过 ognl 表达式创建 Map 集合，并且遍历出集合中的值 <br> <s:set
name="map"
        value="#{'1':'eeeee','2':'ddddd','3':'ccccc','4':'bbbbb','5':'aaaaa'}"></
s:set>
    <s:iterator value="#map" var="o">
        <!--    ${o.key }->${o.value }<br/>   -->
        <!--  <s:property  value="#o.key"/>-><s:property  value="#o.val-
ue"/><br/>  -->
        <s:property value="key" />-><s:property value="value" />
        <br />
    </s:iterator> <br />
    <hr>
4. 通过 ognl 表达式 进行逻辑判断 <br> <s:if

        test="'aa' in {'aaa','bbb'}">
  aa 在 集合 {'aaa','bbb'} 中；
</s:if> <s:else>
  aa 不在 集合 {'aaa','bbb'} 中；
</s:else> <br /> <s:if test="#request.req not in #list">
  不 在 集合 list 中；
</s:if> <s:else>
  在 集合 list 中；
</s:else> <br />
    <hr>
  5. 通过 ognl 表达式的投影功能进行数据筛选 <br> <s:set name="list1"
        value="{1,2,3,4,5}"></s:set> <s:iterator value="#list1.{?#this>2}"
        var="o">
        <!-- #list.{?#this>2}：在 list1 集合迭代的时候，从中筛选出当前
迭代对象 >2 的集合进行显示 -->
```

```
${o }<br />
  </s:iterator> <br />
  <hr>
6. 通过 ognl 表达式 访问某个类的静态方法和值 <br> <s:property
        value="@java.lang.Math@floor(32.56)" /> <s:property
        value="@com.rao.struts2.action.OGNL1Action@aa" /> <br /> <br />
  <hr> 6.ognl 表达式 迭代标签 详细 <br> <s:set name="list2"
        value="{'aa','bb','cc','dd','ee','ff','gg','hh','ii','jj'}"></s:set>
  <table border="1">
        <tr>
              <td>
                    索引
              </td>
              <td>
                    值
              </td>
              <td>
                    奇？
              </td>
              <td>
                    偶？
              </td>

              <td>
                    首？
              </td>
              <td>
                    尾？
              </td>
              <td>
                    当前迭代数量
              </td>
        </tr>
        <s:iterator value="#list2" var="o" status="s">
              <tr bgcolor="<s:if test="#s.even">pink</s:if>">
                    <td>
                          <s:property value="#s.getIndex()" />
                    </td>
```

```
                    <td>
                        <s:property />
                    </td>
                    <td>
                        <s:if test="#s.odd">Y</s:if>
                        <s:else>N</s:else>
                    </td>
                    <td>
                        <s:if test="#s.even">Y</s:if>
                        <s:else>N</s:else>
                    </td>
                    <td>
                        <s:if test="#s.first">Y</s:if>
                        <s:else>N</s:else>
                    </td>
                    <td>
                        <s:if test="#s.isLast()">Y</s:if>
                        <s:else>N</s:else>
                    </td>
                    <td>
                        <s:property value="#s.getCount()" />
                    </td>

                </tr>
            </s:iterator>
        </table> <br>
        <hr>
        7.ognl 表达式 : if/else if/else 详细 <br> <%
    request.setAttribute("aa", 0);
%> <s:if test="#request.aa>=0 && #request.aa<=4">
        在 0-4 之间;
    </s:if> <s:elseif test="#request.aa>=4 && #request.aa<=8">

        在 4-8 之间;
    </s:elseif> <s:else>
        大于 8;
    </s:else> <br>
         <hr>
```

```
                    8.ognl 表达式 : url 详细 <br> <%
        request.setAttribute("aa", "sss");
    %> <s:url action="testAction" namespace="/aa/bb">
                            <s:param name="aa" value="#request.aa"></s:param>
                            <s:param name="id">100</s:param>
                      </s:url> <br /> <s:set name="myurl" value="'http://www.baidu.
com'"></s:set>
                      value 以字符处理: <s:url value="#myurl"></s:url> <br>
                      value 明确指定以 ognl 表达式处理 : <s:url value="%{#myurl}"></
s:url> <br>
                      <hr>
                    9.ognl 表达式 : checkboxlist 详细 <br> 1> .list
                      生成; <br>
                      name:checkboxlist 的名字 <br> list:checkboxlist 要显示的列表 <br>
                      value:checkboxlist 默认被选中的选项 ,checked=checked<br> <s:check-
boxlist
                            name="checkbox1" list="{' 上网 ',' 看书 ',' 爬山 ',' 游泳 ','
唱歌 '}"
                            value="{' 上网 ',' 看书 '}"></s:checkboxlist> <br> 以上生成代
码: <br> <xmp>
                       <input type="checkbox" name="checkbox1" value=" 上网 " id="-
checkbox1-1"
                            checked="checked" /> <label for="checkbox1-1" class="check-
boxLabel">
                            上网
                      </label> <input type="checkbox" name="checkbox1" value=" 看书 "
id="checkbox1-2"
                            checked="checked" /> <label for="checkbox1-2"
class="checkboxLabel">
                            看书
                      </label> <input type="checkbox" name="checkbox1" value=" 爬山 "
id="checkbox1-3" />
                      <label for="checkbox1-3" class="checkboxLabel">
                            爬山
                      </label> <input type="checkbox" name="checkbox1" value=" 游泳 "
id="checkbox1-4" />
                      <label for="checkbox1-4" class="checkboxLabel">
                            游泳
```

```
                    </label> <input type="checkbox" name="checkbox1" value=" 唱歌 "
id="checkbox1-5" />
                    <label for="checkbox1-5" class="checkboxLabel">
                         唱歌
                    </label>" </xmp> 2> .Map
                    生成：<br>
                    name:checkboxlist 的名字 <br> list:checkboxlist 要显示的列表 <br>
                    listKey:checkbox 的 value 的值 <br> listValue:checkbox 的 lablel( 显
示的值 )<br>
                    value:checkboxlist 默认被选中的选项 ,checked=checked<br> <s:check-
boxlist
                         name="checkbox2" list="#{1:' 上网 ',2:' 看书 ',3:' 爬山 ',4:' 游
泳 ',5:' 唱歌 '}"
                         listKey="key" listValue="value" value="{1,2,5}"></s:checkbox-
list> <br>
                    以上生成代码：<br> <xmp> <input type="checkbox" name="check-
box2"
                         value="1" id="checkbox2-1" checked="checked" /> <label
                         for="checkbox2-1" class="checkboxLabel">
                         上网
                    </label> <input type="checkbox" name="checkbox2" value="2" id="-
checkbox2-2"

                         checked="checked"     />     <label     for="checkbox2-2"
class="checkboxLabel">
                         看书
                    </label> <input type="checkbox" name="checkbox2" value="3"
id="checkbox2-3" />
                    <label for="checkbox2-3" class="checkboxLabel">
                         爬山
                    </label> <input type="checkbox" name="checkbox2" value="4"
id="checkbox2-4" />
                    <label for="checkbox2-4" class="checkboxLabel">
                         游泳
                    </label> <input type="checkbox" name="checkbox2" value="5"
id="checkbox2-5"
```

```
                checked="checked"     />     <label     for="checkbox2-5"
class="checkboxLabel">
                唱歌
            </label> </xmp>
            <hr> 10.ognl 表达式 : s:radio 详细 <br> <%
    Sex sex1 = new Sex(1, " 男 ");
    Sex sex2 = new Sex(2, " 女 ");
    List<Sex> list = new ArrayList<Sex>();
    list.add(sex1);
    list.add(sex2);
    request.setAttribute("sexs", list);
  %> 这个与 checkboxlist 差不多;<br> 1>. 如果集合为 javabean:<s:radio name="sex"
                list="#request.sexs"  listKey="id"  listValue="name"></s:radio>
<br>
            2>. 如果集合为 list: <s:radio name="sexList" list="{' 男 ',' 女 '}"></
s:radio> <br>
            3>. 如果集合为 map:<s:radio name="sexMap" list="#{1:' 男 ',2:' 女 '}"
                listKey="key" listValue="value"></s:radio> <br>
            <hr> 11.ognl 表达式 : s:select 详细 <br> 这个与 s:checkboxlist 差不
多;<br>
            1>. 如果集合为 javabean:<s:select name="sex" list="#request.sexs"
                listKey="id" listValue="name"></s:select> <br> 2>. 如果集合为
list:<s:select
                name="sexList" list="{' 男 ',' 女 '}"></s:select> <br> 3>. 如果集
合为 map:<s:select
                name="sexMap" list="#{1:' 男 ',2:' 女 '}" listKey="key" listVal-
ue="value"></s:select>

            <br> 到此主要的 ognl</SPAN> 标签已经介绍完毕 ... 由于表单标
签相对简单不介绍了 ....
    </body>
  </html>
```

其中使用到的 JavaBeans 代码如示例代码 7-31 所示。

示例代码 7-31　JavaBean 代码

```
package com.xtgj.struts2.domain;
public class Sex {
    private int id;
    private String value;
    public Sex() {
    }
    public Sex(int id, String value) {
        this.id = id;
        this.value = value;
    }
    public int getId() {
        return id;
    }
    public void setId(int id) {
        this.id = id;
    }
    public String getValue() {
        return value;
    }
    public void setValue(String value) {
        this.value = value;
    }
}
```

运行结果如图 7-2 所示。

图 7-2 运行结果

7.3 作业

1. 根据本章示例练习 Struts 2 标签的应用。
2. 利用 Struts 2 标签完成用户注册页面的构建。

第 8 章　Struts 2 国际化

本阶段目标

完成本章内容后，你将能够学到：

- ✧ 全局国际化配置。
- ✧ 包范围国际化配置。
- ✧ Action 范围国际化配置。

本阶段给出的步骤全面详细，请学员按照给出的上机步骤独立完成上机练习，以达到要求的学习目标。请认真完成下列步骤。

8.1　指导（1 小时 10 分钟）

国际化是商业系统中不可或缺的一部分，所以无论学习的是什么 Web 框架，它都是必须掌握的技能。利用 Struts 的优势，只需要编写不同语种的属性文件，就可以轻松实现多语种功能，满足企业国际化的需要。

8.2　练习（50 分钟）

8.2.1　国际化的示例

下面的案例详细介绍了全局国际化，包范围国际化和 action 范围国际化的应用。根据提示理解国际化的含义和应用，并实践。

第一步，定义全局国际化资源文件， ApplicationResources_en_US.properties 代码如示例代码 8-1 所示。

示例代码 8-1 ApplicationResources_en_US.properties

```
hello.jsp.title={0},Hello-A first struts program,{1}
hello.jsp.page.heading=Hello World\!A first struts application
hello.jsp.prompt.person=Please enter a UserName to sag Hello to \:
hello.jsp.page.hello=Hello
hello.jsp.button.submit=-\=Submit\=-
hello.jsp.button.reset=-\=Reset\=-
hello.jsp.img.src=/stuts-power.gif
hello.jsp.img.alt=powered by struts
hello.dont.talk.to.monster=We don't want to say hello to Monster\!\!\!
hello.no.username.error=Please enter a <i>UserName</i> to say hello to\!
```

ApplicationResources_zh_CN.properties 代码如示例代码 8-2 所示。

示例代码 8-2 ApplicationResources_zh_CN.properties

```
hello.jsp.title={0},Hello- 一个 struts 程序 ,{1}
hello.jsp.page.heading=Hello World\! 一个 struts 应用程序
hello.jsp.prompt.person= 请输入你要问候的用户名:
hello.jsp.page.hello= 你好
hello.jsp.button.submit=-= 提交 =-
hello.jsp.button.reset=-= 重置 =-
hello.jsp.img.src=/stuts-power.gif
hello.jsp.img.alt= 由 struts 组织支持
hello.dont.talk.to.monster= 我们不想问候 Monster!
hello.no.username.error= 请输入一个 <i> 用户名 </i> 并向他问候！
```

第二步，将两个全局资源文件拷贝至 src 根目录下。

第三步，在 struts.xml 中配置国际化常量信息，struts.xml 代码如示例代码 8-3 所示。

示例代码 8-3 struts.xml

```
<?xml version="1.0" encoding="UTF-8"?>
<!DOCTYPE struts PUBLIC
    "-//Apache Software Foundation//DTD Struts Configuration 2.3//EN"
    "http://struts.apache.org/dtds/struts-2.3.dtd">
<struts>
    <constant name="struts.custom.i18n.resources"
        value="ApplicationResources" />
```

```
    <package name="user" namespace="/user" extends="struts-default">
        <global-results>
            <result name="msg">msg.jsp</result>
        </global-results>
        <action name="user_*" class="user.UserAction" method="{1}">
            <result name="success">success.jsp</result>
        </action>
    </package>
</struts>
```

第四步，定义 action 类 UserAction.java，其代码如示例代码 8-4 所示。

示例代码 8-4　UserAction 类

```
package user;
import com.opensymphony.xwork2.ActionContext;
import com.opensymphony.xwork2.ActionSupport;
public class UserAction extends ActionSupport {
    public String read() {
        String text = this.getText("hello.jsp.title",new String[]{"Welcome","Good Luck!"});
        ActionContext.getContext().put("text", text);
        return "success";
    }
}
```

第五步，定义 JSP 视图 i18n.jsp，其代码如示例代码 8-5 所示。

示例代码 8-5　i18n.jsp

```
<%@ page language="java" import="java.util.*" pageEncoding="UTF-8"%>
<%@taglib prefix="s" uri="/struts-tags"%>
<!DOCTYPE HTML PUBLIC "-//W3C//DTD HTML 4.01 Transitional//EN">
<html>
    <head>
        <title>My JSP 'i18n.jsp' starting page</title>
    </head>
    <body>
        <s:text name="hello.jsp.title">
```

```
                <s:param>Welcome</s:param>
                <s:param>Good Luck!</s:param>
            </s:text>
            <br />
        </body>
    </html>
```

注意，该页面应存放至 WebRoot 目录的 user 文件夹下。

在浏览器中输入 http://localhost:8080/I18N/user/i18n.jsp，运行结果如图 8-1 所示。

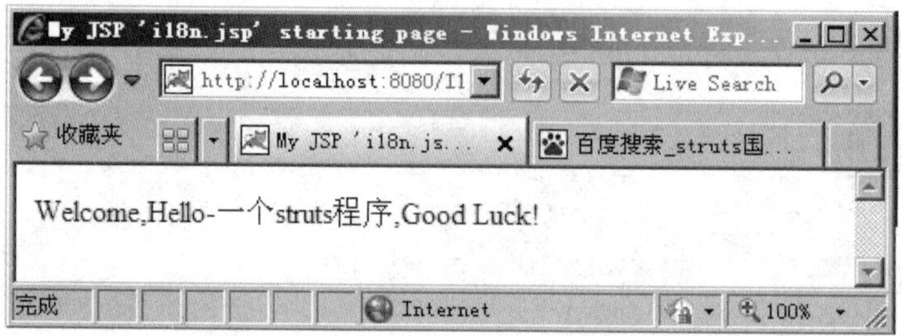

图 8-1 运行结果

第六步，定义包 user，在该包下存放两个国际化资源文件，package_en_US.properties 代码如示例代码 8-6 所示。

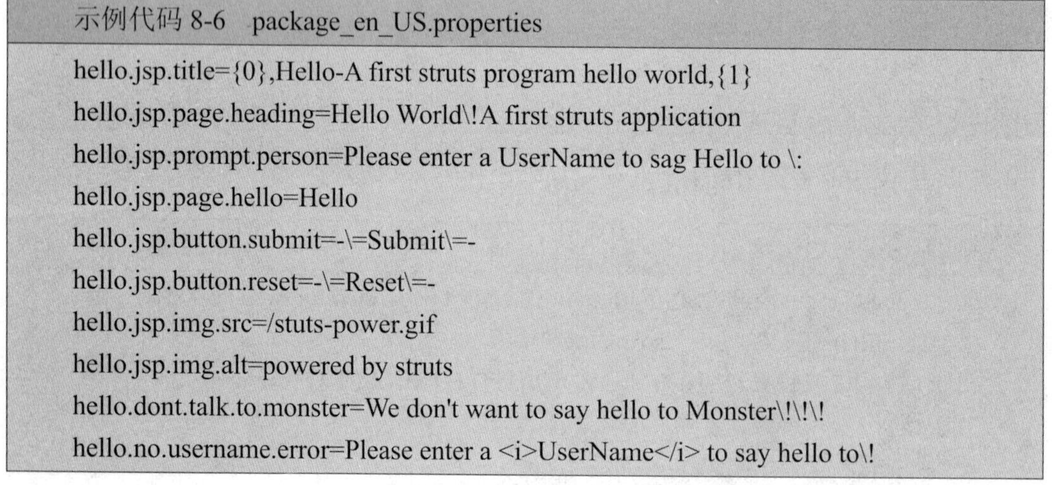

示例代码 8-6 package_en_US.properties

```
hello.jsp.title={0},Hello-A first struts program hello world,{1}
hello.jsp.page.heading=Hello World\!A first struts application
hello.jsp.prompt.person=Please enter a UserName to sag Hello to \:
hello.jsp.page.hello=Hello
hello.jsp.button.submit=-\=Submit\=-
hello.jsp.button.reset=-\=Reset\=-
hello.jsp.img.src=/stuts-power.gif
hello.jsp.img.alt=powered by struts
hello.dont.talk.to.monster=We don't want to say hello to Monster\!\!\!
hello.no.username.error=Please enter a <i>UserName</i> to say hello to\!
```

package_zh_CN.properties 代码如示例代码 8-7 所示。

示例代码 8-7　package_zh_CN.properties

```
hello.jsp.title={0},Hello- 一个 struts 程序 ,hello world,{1}
hello.jsp.page.heading=Hello World\! 一个 struts 应用程序
hello.jsp.prompt.person= 请输入你要问候的用户名：
hello.jsp.page.hello= 你好
hello.jsp.button.submit=-= 提交 =-
hello.jsp.button.reset=-= 重置 =-
hello.jsp.img.src=/stuts-power.gif
hello.jsp.img.alt= 由 struts 组织支持
hello.dont.talk.to.monster= 我们不想问候 Monster!
hello.no.username.error= 请输入一个 <i> 用户名 </i> 并向他问候！
```

第七步，在 i18n.jsp 填入如示例代码 8-8 所示代码片段。

示例代码 8-8　在 i18n.jsp 填入代码片段

```
<s:i18n name="user/package">
    <s:text name="hello.jsp.title">
        <s:param>Welcome</s:param>
        <s:param>Good Luck!</s:param>
    </s:text>
</s:i18n>
<br />
```

在浏览器中输入 http://localhost:8080/I18N/user/i18n.jsp，运行结果如图 8-2 所示。

图 8-2　运行结果

第八步，在 user 包下定义 Action 类，UserAction.java，其代码同第四步。

第九步，配置 struts.xml，其代码同第三步。

第十步，在 user 包下添加国际化资源文件，UserAction_en_US.properties 其代码如示例代码 8-9 所示。

示例代码 8-9 UserAction_en_US.properties

```
hello.jsp.title={0},Hello-A first struts program hello China,{1}
hello.jsp.page.heading=Hello World\!A first struts application
hello.jsp.prompt.person=Please enter a UserName to sag Hello to \:
hello.jsp.page.hello=Hello
hello.jsp.button.submit=-\=Submit\=-
hello.jsp.button.reset=-\=Reset\=-
hello.jsp.img.src=/stuts-power.gif
hello.jsp.img.alt=powered by struts
hello.dont.talk.to.monster=We don't want to say hello to Monster\!\!\!
hello.no.username.error=Please enter a <i>UserName</i> to say hello to\!
```

UserAction_zh_CN.properties 代码如示例代码 8-10 所示。

示例代码 8-10 UserAction_zh_CN.properties

```
hello.jsp.title={0},Hello- 一个 struts 程序 ,hello China,{1}
hello.jsp.page.heading=Hello World\! 一个 struts 应用程序
hello.jsp.prompt.person= 请输入你要问候的用户名:
hello.jsp.page.hello= 你好
hello.jsp.button.submit=-= 提交 =-
hello.jsp.button.reset=-= 重置 =-
hello.jsp.img.src=/stuts-power.gif
hello.jsp.img.alt= 由 struts 组织支持
hello.dont.talk.to.monster= 我们不想问候 Monster!
hello.no.username.error= 请输入一个 <i> 用户名 </i> 并向他问候！
```

第十一步，在 i18n.jsp 填入如示例代码 8-11 所示代码片段。

示例代码 8-11 在 i18n.jsp 填入代码片段

```
<a href="user_read.action"> 读取国际化信息 </a>
```

第十二步，定义 JSP 视图 success.jsp，其代码如示例代码 8-12 所示。

示例代码 8-12 success.jsp

```
<%@ page language="java" import="java.util.*" pageEncoding="UTF-8"%>
<!DOCTYPE HTML PUBLIC "-//W3C//DTD HTML 4.01 Transitional//EN">
<html>
    <head>
        <title>My JSP 'success.jsp' starting page</title>
```

```
        </head>
        <body>
            读取成功:
            <br>
            信息:${text }
        </body>
    </html>
```

在浏览器中输入 http://localhost:8080/I18N/user/i18n.jsp，点击页面上的链接后，运行结果如图 8-3 所示。

图 8-3　运行结果

8.3　作业

1. 根据本章示例练习 Struts 2 国际化应用。
2. 利用 Struts 2 国际化完成用户注册页面的构建。